T0234275

Expedition into the Nanoworld

Alberto Diaspro

Expedition into the Nanoworld

An Exciting Voyage from Optical Microscopy to Nanoscopy

 Springer

Alberto Diaspro
Department of Physics
Istituto Italiano di Tecnologia
University of Genoa
Genoa, Italy

ISBN 978-3-030-94471-1   ISBN 978-3-030-94472-8   (eBook)
https://doi.org/10.1007/978-3-030-94472-8

This Springer imprint is published by the registered company Springer Nature
Switzerland AG
The registered company address is: Gewerbestrasse 11, 6330 Cham, Switzerland

Faust'o Happy Is very
verde.
Because, if you pperd…e
Now, cche vuoi? Perchè
sei Ciccio e nun ssei
Ggiuda!
Con le Bermuda, nun fa
accussi.
(F. Masala, E. Zamponi, Argentario, 1982)

Dedicated to Teresa.

Foreword

"Alberto Diaspro takes the reader on a kaleidoscopic journey of microscopy as colourful and full of beauty as science and life itself."

Stefan W. Hell
2014 Nobel Laureate "for the development of microscopy in super-resolved fluorescence."

"It is an interesting period for quality scientific divulgation in our country and we can consider ourselves lucky, given the thick curtain of misinformation that generally obscures the citizens of the planet and of Italy in particular. That's why I like to greet this popular text that has, however, much more (and, if I may, better) than other texts. In the sense that, often, scientists describe their experiences in an aseptic way and focus, as it is right, on the general goal of their effort: to explain a concept or, even, the cosmos, no matter if macro or micro. In this case, Diaspro continually and very pleasantly mixes personal, cultural and scientific topics and themes in an admirable way, even having fun on the border of humanism and science. Contributing, in my opinion, to the reunification of what are just two sides of the same general culture, that of *sapiens*. In a continuous change of focus, never so pertinent to the subject matter, the reader enjoys himself and learns concepts, and even curiosities, that make that knowledge interesting and in some way indispensable. I want to say that, after reading, one will be surprised not to have known about those topics before and will want to know even more. A sure testimony that the mission of divulgation is accomplished."

Mario Tozzi
Italian geologist, popularizer of science
and essayist, author and TV personality
Max Planck Institute
for Biophysical Chemistry
Goettingen, DE

Acknowledgements

After the war... there was a desire to dance that made light!

F. Guccini to D. Bianchi,
Propaganda Live, La 7 broadcast, December 6th, 2019

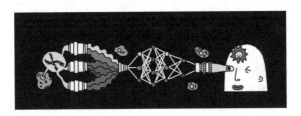

Thanks to Marta Fioravanti for her graphical representation of this book thinking about MOMIX—Graphic donated by Marta Fioravanti - https://www.instagram.com/nonlineare/ (2020).

The "thank you" are so many, so many that they almost have to be recited, chanted aloud, because it is so, because it is right. Here they are in a row, in order, a somewhat disordered order, perhaps better to say by blocks.

Thanks to Teresa, Claudia and Irene, the most important women in my life. Thanks to little Sissi who kept me company in my morning writing at my desk. Thanks to Francesco Avola, more than a son-in-law. Thanks to Anna Mercurio, Mario Arace, Carmen Gloria and Mario Cassano, special people. Thanks to Bruno, whose student I was, and Franca Bianco for welcoming me and giving me so much more along with Betta, Francesca and Pietro. Thanks to my father, Renzo, and Franca for something they know. Thanks to Maurizio Caviglia for his way of supporting new ideas. Thanks to Antonio Borsellino for spurring me on when I wanted to stop. Thanks to Giuliano Toraldo di Francia for his lessons. Thanks to Carlo Tacchetti for allowing me to continue as a microscopist in the most difficult moments. Thanks to Enrico Gratton who is my model, I realize unequalled. Thanks to the mythical E of Cassini for being always present. Thanks to Francesco Guccini, whose songs have accompanied my studies, my research and the writing of this book. Thanks to Raffaella Zuccari who convinced Francesco Guccini to follow a forty-minute presentation on multiphoton microscopy at Via Paolo Fabbri 43. Thanks to Guendalina Zuccari and Michel Dabove for the table in Sassello and more. Thanks to Roberto Cingolani who has believed in my project on optical nanoscopy at the Italian Institute of Technology in Morego. Thanks to Paolo Bianchini, Valentina Caorsi, Davide Mazza, Giuseppe Vicidomini, Ilaria Testa, Francesca Cella Zanacchi and Emiliano Ronzitti with whom I was able to start building a group of world-renowned microscopists at the Department of Physics of the University of Genoa in Via Dodecaneso 33 between L303 and L308. A double thanks for Paolo Bianchini for sharing with me microscopy adventures... "lifetime". Thanks to the Biophysicists of L303 in the

Department of Physics and in particular to Alessandra Gliozzi, Ranieri Rolandi, Mauro Robello and Alberto Blasi, who welcomed me to Physics on July 3, 1995. Thanks to Marco Castello and Simonluca Piazza for their patient reading of the first draft and much more. Thanks to Jenu Chacko for the artificial and natural intelligence. Thanks to Michele Bertolotto, Giuseppe Cittadini, Emilio Di Maria and Giovanni Caruso with whom I started the adventure of Biophysics at the Institute of Biophysics of the Uni-versity of Genoa in Viale Benedetto XV. Thanks to Ro Kampman for Biophysics. Thanks to Ammasi Periasamy, Karsten Konig and Peter T. C. So for being an unfailing trio in my life. Thanks to Grazia Tagliafierro, Paola Ramoino, Cesare Usai, Laura Vergani and Paolo Germano for microscopy together. Thanks to Gianna Castellaro with whom I shared an atypical dissertation on microscopy. Thanks to Carlo Castellano for my first professional contract on polarized light. Thanks to Mario Faretta, Pier Giuseppe Pelicci and Pier Paolo di Fiore for microscopy applied to real oncology. Thanks to Giacomo Torre, Giorgio Cuttica, Giacomo Pasini, Giuliano Armano, Giancarlo Parodi and Giulio Peirone for making me an electronic engineer. Thanks to "Bambi" Giovanni Adorni, Agostino Bozano and Balbina Scano for the first house where I lived in Salita Carbonara 57/2 above the legendary Vito. Thanks to all the Ph.D. students, young researchers, and colleagues with whom I have inter-acted over the years. Among them a special thanks to Pio Benedetti, Mario Bertero, Luca Lanzanò, Ranieri Bizzarri and Giberto Chirico, remembering Sabrina Beretta.

Thanks to Tulle Hazzelrigg and Martin Chalfie for their continued teachings in science and lifestyle.

Thanks to Erwin Neher and Ada Yonath for the discussions and teachings.

I was lucky for the friendship with Mats Gustafsson.

Among the many microscopists special thanks to Colin J. R. Sheppard, Peter Saggau, Carlos Bustamante, Eric Betzig, Stefan W. Hell, Joerg Enderlein, Rainer Heintzman, Tony Wilson, Fred Brakenhoff, Sara Abramsshon, Niek van Hulst, Maria Garcia Parajo, Paula Sampaio, Fred Wouters, Gertrude Bunt, Osamu Nakamura, Satoshi Kawata, Fu Jen Kao, Peter Torok, Rimas Juskaitis, Fedja Bobanovic and many others.

Thanks to Cristiana Ricci, Paolo Sapuppo, Massimo Scauso, Claudio Clementi, Ryu Nakamura, Rolf Borlinghaus, Irmi Steinmetz, Vanessa Lurquin, Kees van Oord, Peter Drent, Hans van der Voort and Gitta Hamel. Thanks to Michael Stanley for his friendship and the custom-made, for me, technical filters to start with two-photon excitation microscopy. Thanks to Ettore Castiglioni and Martin Hoppe. Thanks to Cristiano Viappiani and Stefania Abbruzzetti for the way they do science and life. Thanks to Elena Bauer for a beautiful idea and more. Thanks to Samantha Cristoforetti for letting me enter, virtually, the orbiting space station.

Special thanks for the discussions of science, life and philosophy to Toshiyuki Masai, a great professional and dear friend.

Thanks to Gianni Berengo Gardin and Caterina Stiffoni for pushing me to see inside photographs more than what they "normally" show.

Thanks to Franca Speranza and Andrea Cosulich for their way of stimulating cultural and scientific discussion.

Thanks to Nicla Vassallo more than a professor, more than a philosopher… a unique friend to discuss with about everything.

Thanks to Books, paper or digital, for that consultation that, since childhood, has seen the Encyclopedia Treccani opened for every question. An old habit never abandoned.

Thanks to the Internet and Wikipedia, which even within the limits of the flow of information on the net offer the possibility of acquiring information critically. It is up to us to be lazy or not. Even the Treccani is on the net.

Thanks to Music, all of it, on vinyl or CD or on the net that has always accompanied my reading or studying and my writing.

Thanks to Franz DiCioccio, excellent drummer of the Premiata Forneria Marconi (PFM), a "multimodal" musician for culture all-around. Thanks to Patrick Dijvas, PFM, a great jazz man. Thanks to Iaia De Capitani and Giovanni Caprara for making possible the jam session with the microscope. Thanks to Massimo Sideri for believing in the route from microscopy to nanoscopy.

Thanks to Pietro Greco, Rossella Panarese and Gianfranco Bangone for their lessons in communicating Science.

Thanks to Ulrico Hoepli (1847–1935), a patron, who contributed to cultural development by founding the "Biblioteca Popolare Ulrico Hoepli" in 1921, and in 1930 he donated the planetarium to the city of Milan. Thank you for that idea of "supporting institutions and initiatives of public utility or that work to promote the sciences and the arts". Thank you to those I have not mentioned but who are very much in my mind.

Thanks to you readers who have endured pages with formulas that were too simplified for some and too difficult for others, but then you got there, you got to the bottom.

Thanks for the dance! (L. Cohen, 2019)

About This Book

Some formulas and concepts that may be difficult to understand are used in the discussion. This is done intentionally by asking the reader to make a small effort to maintain rigor in a treatment that is intended to be "popular". The suggestion is to read the formulas as text, like the text of a poem. The formulas are not isolated or numbered, intentionally. Similarly, concepts that are more difficult to understand are not isolated in a box. The author's idea is that they should be harmonious with the rest of the text. After all, they help to maintain the "narrative time" of the story.

Contents

1

Curious Premise

*Tall, solemn, dressed in black, it seemed to me to see Granny Lucia
again; [...] Oh Granny, Oh Granny! oh how beautiful she was
when I was a child! tell me again, tell this wise man the news of her
who seeks her lost love! [...] You sleep to my desperate cries, and the
cock crows, and you won't wake up.*
Giosuè Carducci, *Davanti a San Guido*, in *Rime nuove* (1906).

My grandmother Anna was a beautiful woman (Fig. 1.1).
I was a curious and lively child like all my peers. When she
died of cancer, an impossible promise rose from my heart
and I whispered to her, "I'll understand why, Grandma
dear, so you won't die anymore." It sounds a crazy promise, the next will show it was not so crazy. I didn't realize that I already was a young microscopist, at that time. I
was born in Genoa, but I had spent most of my childhood
in Verona. It often snowed and those flakes that covered
everything in white fascinated me. One day in my stamp
collection, among those detached from a blue envelope, I

© Casa Editrice Ulrico Hoepli S.p.A. 2022
A. Diaspro, *Expedition into the Nanoworld*,
https://doi.org/10.1007/978-3-030-94472-8_1

Fig. 1.1 Grandmother Anna Mercurio in Arace

found a coloured stamp with a very white and beautiful snowflake in the center (Fig. 1.2).

It was then that I realized that there are things that the eyes are not able to see, besides Santa Claus and his nine reindeers. The snowflakes, on those "Santa Lucia" days when everything was quickly covered in white around me, did not appear like that postage stamp. When I let them rest on the palm of my hand, the newly captured snowflakes became transparent and transformed into drops of water. Because of those strange combinations of life, I was

Fig. 1.2 The snowflake

given a small microscope with a cedar wood box containing slides in that winter.

However, I had ten rectangular slides on which, protected by a thinner square of glass, had been deposited the strangest things: from butterfly wings to cat hairs, from leaf veins to insect legs or thistle roots. In the package there were also five empty slides to give space to the creativity of budding microscopists. Creativity was mainly developed in finding something attractive to observe: a piece of rotten banana, a bit of spit or a few drops of blood. Stuff from Tom Sawyer and Huckleberry Finn, young boys protagonists of daring stories (Mark Twain, 1876 and 1884) or from Hardy Boys and Nancy Drew, very young amateur detectives ripped from their books and gathered in a single telefilm by Universal TV between 1977 and 1979.

There, I was ready to observe the snowflakes. Once the flakes had been captured, it was a question of orienting a small mirror that collected the light of the day and sent it to the slide on which I had dropped the snow so that I could observe it with that curved piece of glass, the lens, which made it possible to form the image observable to the eye. Focusing was fast, I was skilled in using the knobs of the microscope. I had to be fast before snowflakes could turn into water. In fact, all these operations took time and, in the meantime, the snow was melting. I learned that it was fundamental to cool the slide, perhaps leaving it immersed in the snow in the garden while I was at school. The garden of my house in via Fiume 2, the one with a beautiful willow tree in the middle, was my laboratory (Fig. 1.3).

So, conducting measurements in the cold, being quick in manual operations and knowing that the subject to observe could change over time were the first three things I learned as a promising microscopist. At home, after all, I did not have all these problems with the preparations in the cedar box. I would bring the microscope up to temperature along with the slides by hiding it under the willow tree, so that when I returned from school, before going up to the house, I could try to observe the snowflakes. The reasons for my delay in going up to the house were immediately revealed by the wet trousers at knee height on which I leaned, immersed in the snow during the observations.

The passion for the microscope remained intact over the years becoming one of the staples in my university and research career. During the third year of electrical engineering at the University of Genoa I was struck by a lesson on electromagnetic fields given by Professor Bruno Bianco. Professor Bianco explained, starting from Maxwell's equations that were the cross and delight of student engineers,

Fig. 1.3 Via Fiume 2 in Verona, the old willow tree and the bench. Drawing by Gianluigi and Lucio Perin, Verona, 6-5-1968

how visible light was an electromagnetic wave of which we could "see" the squared modulus with our eyes as with a video camera and how it was in a very restricted portion of the spectrum of energies carried by electromagnetic waves. Then, in an exciting double somersault, he explained that what was happening in space could be treated like music, with harmonics and spatial frequencies. It could be treated with high, medium and low frequencies, the same ones I used to set on my amplifier at home when I switched from the Italian singer-songwriter Francesco Guccini to the progressive sound of Pink Floyd. Call them sounds, vibrations, or images, but they can all be described in a harmonic way as nature is after all. A well detailed image of a thick spider web can be thought of as an image of high frequencies, spatial rather than temporal or musical. The high frequency sound of a violin. A beautiful photograph of a landscape with a sunset will contain mostly medium frequencies, the discreet sound of a classical guitar. While the final scene in the fog of *Casablanca*, with its "Today perhaps we inaugurate a beautiful friendship", is dominated by low frequencies, like Jacqueline Du Pré's cello in the first movement of Elgar's cello concerto. It was called Fourier optics, spatial frequencies instead of distances. There was no mathematical constraint on treating images with the Fourier transform (Fig. 1.4), which was permissible for any phenomenon occurring in nature.[1]

At that time, I only had optical microscopy in my thoughts as a potential future researcher, that its transforming real images into microscope images that I could process in the so called Fourier domain, the subject of a book by Joseph W. Goodman that was the object of my first professional purchase, made with the proceeds of my first salary.[2]

What the eyes do not see of living things are details, because they are too fine, and molecules, because biological molecules, for the most part, do not absorb visible radiation, so they are unable to produce contrast. Biological molecules behave more or less like a crystal glass. Visible light, just think of it as the electromagnetic wave that propagates, enters the crystal glass and exits unaltered in amplitude and slowed down in time. We catch the intensity but not the slowing down, since the speed of light is quite high. It was determined by Ole Rømer, Galilei's disciple, in 1676 while he was working at the royal observatory in Paris directed by Gian Domenico Cassini, who together with Robert Hooke in 1664

---

[1] R.N. Bracewell, The Fourier Transform and Its Applications. (McGraw-Hill, 1986).

[2] J.W. Goodman, *Introduction to Fourier Optics, 3rd ed.*, Roberts & Company, 2005 (McGraw Hill, 1968).

Fig. 1.4 An image and its Fourier transform in two dimensions. The diagonal momentum to the right is "flipped" in the transform at the top left. The central bright spot, coordinate 0-0, collects all the "energies" in play distributed over the various frequencies

discovered Jupiter's Great Red Spot and, in about 1690, was the first to observe the differential rotation of Jupiter's atmosphere.

The fact that biological molecules do not absorb or absorb very little light in the visible region of the electromagnetic spectrum (Fig. 1.5) is reasonable if you think that we are under visible light for most of the day. If the radiation were absorbed, we would be "cooked like chickens on a spit". As soon as we move out from the visible region, for example into the ultraviolet, the radiation can be absorbed. Molecules such as DNA and a large part of proteins absorb energy in the ultraviolet and the consequent rise in temperature could deteriorate them. The sun, being outdoors, is pleasant but, as always, excessive exposure also exposes us to ultraviolet radiation which, when

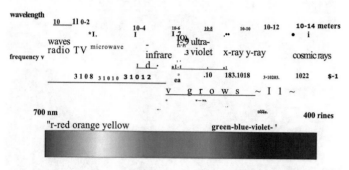

Fig. 1.5 The visible light of the rainbow is a small portion of the waves that "envelop" us. Increasing wavelength indicates lower energies, frequency the opposite

no longer filtered by atmospheric pollution, can damage DNA and proteins.

The other reason why the eyes can't see is because of the fineness of the detail, you might say "things are too small", but in fact it's the small distances between objects that sometimes don't allow you to consider them as two separate things. How close are two proteins or two elephants? One meter, one micrometer or one billionth of a meter, or the nanometer, which is like dividing a hair into one hundred thousand parts. The eye can see a hair well, but it loses detail if it splits into too many parts (Fig. 1.6).

The optical microscope, therefore, makes it possible to see the doubly invisible using a curved piece of glass, the lens, and the rainbow, the source of illumination to which matter reacts by sending, among other things, light signals.

From the snow to the lessons of Bruno Bianco my research started and continued until today, until the time of super resolution and what I now call multi-message microscopy based on MOMIX (multimodal optical microscopy image correlation sensing). We have arrived at

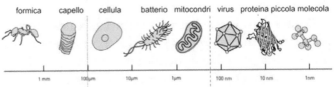

| formica | capello | cellula | batterio | mitocondri | virus | proteina piccola molecola |

| 1 mm | 100μm | 10μm | 1μm | 100 nm | 10 nm | 1nm |

Fig. 1.6 Our eye can see the hair but with the light microscope we can observe the cells in detail, not really see the single proteins unless we have "barbs" (*I Barbapapà*, 1970)

the time of multimodal and liquid microscopy, *liquitopy*, through the many stories of microscopy that have seen and see as protagonist just a piece of curved glass and rainbow light.

It is a matter of fact that one of these stories makes me particularly proud.

It is a very simple story, a scientific collaboration that began when a researcher from Padua involved me in research on neuroblastoma, a terrible disease that affects children. A solid tumor of young age, neuroblastoma is responsible for 15% of deaths attributed to childhood neoplasms and in 50% of cases at the time of diagnosis it is already metastatic and stubborn to chemotherapy.[3] She had been a victim: her son. In about two years we were able to demonstrate that aloe-emodin, a molecule derived from the aloe plant, could be useful in that battle for life.

---

[3] J.M. Maris et al., Chromosome 6p22 locus associated with clinically aggressive neuroblastoma, "The New England Journal of Medicine," June 12, 2008, 358:2585–2593, https://doi.org/10.1056/NEJMoa0708698.

We were able to observe its action thanks to the fact that it is an anthraquinone visible under the novel "two-photon" microscope, built in my lab at the Department of Physics, because of its luminescence, present in the parenchyma of the leaves of some species of aloe.

What I am still most proud of is not so much the work published in the prestigious journal Cancer Research,[4] but the fact that my colleague decided to have more children.

So, a story of microscopy, one of the many stories of microscopy, a story between a piece of glass and the light of the rainbow, became an incredible story of love of life and for life.

Look at the sky. "The sky is blue because you want to know why the sky is blue." (Jack Kerouac, *The Dharma Wanderers*, 1958). Ask yourself, "Did the sheep eat the flower or not?" (Antoine de Saint-Exupéry, *The Little Prince*, 1943). And you will see how everything changes!

That's microscopy, baby!

---

[4] T. Pecere et al., Aloe-emodin Is a New Type of Anticancer Agent with Selective Activity against Neuroectodermal Tumors, "Cancer Research," 60 (11):2800 2804, 2000.

2

Just Observe!

*But how I wish I had your eyes, wide open to the world like blotting
papers - and your laughter clean and full, almost without sorrow
or repentance, - but how I wish I had everything still to look at
like books to leaf through,-and still have everything, or almost
everything, to try...*
Francesco Guccini, *Culodritto*, in "Signora Bovary", EMI music,
(1987)

How can I see what's going on inside a cell? Understand
how the molecules are organized to make it into a lung or
a hair? Can you predict if a cell will get sick, if we will get
sick? If so, could we be cured? The answer is yes. Let's try
to start here.

What do an aphorism by American baseball player
Yogi Berra (1925–2015) and the legendary lecture given
at Caltech in 1959 by Richard Feynman (1918–1988),
Nobel Laureate in physics in 1965, have in common? "Just
observe!" they both exhort (Fig. 2.1).

© Casa Editrice Ulrico Hoepli S.p.A. 2022
A. Diaspro, *Expedition into the Nanoworld*,
https://doi.org/10.1007/978-3-030-94472-8_2

Fig. 2.1 Richard Feynman, theoretical physicist, and Yogi Berra, catcher for the New York Yankees

For Yogi Berra, catcher for the New York Yankees, observing even the smallest, almost invisible, details in the pitcher's movements were the key to intercepting the ball and scoring. This was the key for his success, an incredible career between 1946 and 1965, winning more than Joe Di Maggio.

"You can observe a lot by watching!" was one of his most famous aphorisms, known in the USA and around the world as "yogisms".[1]

The observation is also one of the key points of the memorable lecture given by the American physicist Richard P. Feynman at the California Institute of Technology in Pasadena, the Caltech, temple of

---

[1] Y. Berra, T. Horton, Yogi. It Ain't Over, McGraw-Hill, 1989.

Fig. 2.2 An atomic electron fence. An enclosure of 48 iron atoms on a smooth copper surface confines electrons on the copper surface. The electrons are not balls going around something but "ripples"

technology, in December 1959. That presentation marked the birth of nanotechnology.[2]

Nanotechnology refers to our ability to observe and control matter, made up of atoms, on a very small scale (Fig. 2.2). Feynman, who was awarded the Nobel Prize for Physics in 1965 for his work on quantum electrodynamics, pointed to microscopy and biological molecules as two fixed points for the advancement of all sciences and technology, as has happened and is happening.

Feynman began roughly like this:

---

[2] *There's Plenty of Room at the Bottom: An Invitation to Enter a New Field of Physics*, speech at the annual meeting of the American Physical Society at Caltech on 29 December 1959: R.P. Feynman, *Il piacere di scoprire*, Adelphi, Milan 2002.

I would like to illustrate to you a field of physics in which little has been done so far, but which, on a theoretical level, offers enormous potential. It's a rather peculiar field but it could help us understand many interesting things about strange phenomena that occur in complex situations, and, above all, it could have countless technological applications. What I want to talk about is the problem of manipulating and controlling things at a small scale.

This changed everything and brought to the forefront a subject from which to learn, which offered and offers all-round insights into science, literature and the arts. As Primo Levi (1919–1987)—elected as the greatest science writer of all time by the English Royal Institution—stated: "I have often set foot on the bridges that connect scientific and literary culture, climbing over a crevasse that has always seemed absurd to me". Please, consider that at the end, the subject is the living. After all, we, 7 billion people with about 76 organs each and 100 thousand billion cells that preserve and use two meters of DNA crammed into a few millionths of a meter are an excellent example of the "multicultural" functioning of a nanomachine at the macro scale (Fig. 2.3).

For some reasons, our eyes cannot see all this at the small scale, at the nanoscale. The understanding of the functioning and the evolution of those mechanisms that make living things live and that can make us progress occurs precisely at that scale. To understand what happens, for Richard Feynman as for Yogi Berra "it would be enough to observe. Just observe!".

Feynman had made it clear that it was necessary to improve the electron microscope further to access the secrets of the living.

I like to think of it as a heart-in-the-throat, breathtaking reading of the passage where he considers DNA,

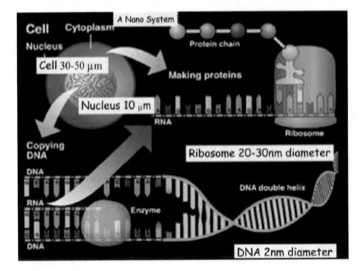

Fig. 2.3 The biological cell is a nano-machine; events on the scale of DNA, 2 nm, produce consequences on the cell, tissues, and organs of living things

proteins, and the microscope for the observation process "There's a plenty of room at the bottom."[3]

Feynman was particularly interested in the possibilities of making increasingly dense and compact computer circuits and microscopes that would allow objects even smaller than those visible through an electron microscope to be observed.

The latter possibility was realized in the mid-1980s with the invention of the first tunnel-effect microscope and then the atomic force microscope (Fig. 2.4).[4] Today, the optical microscope goes as far as the nanoscale without magic, tricks, and without violating the laws of physics.

---

[3] R.P. Feynman, *ibid.*

[4] A. Diaspro, R. Rolandi, *Scanning force microscopy for imaging biostructures at high-resolution*, "European Journal of Histochemistry," 41(1):7–16, 1997.

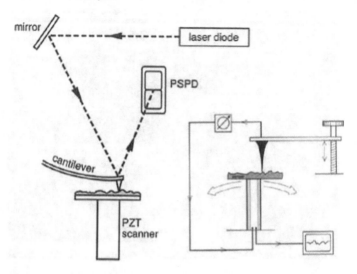

Fig. 2.4 To make an atomic map you can use the mechanism of an "old record player". The stylus is attracted or repelled according to the distance from the surface and by measuring the forces of traction or repulsion we can make maps with atomic detail

What the eyes see and don't see tells us what, and how much, information we are able to gather and process, in our case with the microscope. DNA and biological molecules are incredibly efficient systems for that control of matter at a very small scale. Feynman observed that for biologists…

> The fact that an incredibly small space could contain enormous amounts of information is nothing new.

A lot of information is contained in a very small fragment of our cells, in a chain of DNA molecules, in which about 50 atoms are used for each "bit" of information about that cell. All the information—whether we have green or brown eyes, whether we think or not, or the fact that in the embryo the jaw must develop with a small hole on the side so that a nerve, later, can pass through it as it grows.[5] So, making microscopes more powerful in capturing details, molecule by molecule, could provide answers to crucial questions in biology and beyond; it was 1959. This was Feynman's incitement.

> What is the base sequence of DNA? What happens when a mutation occurs? What is the relationship between the order of bases in DNA and the order of amino acids in proteins? What is the structure of RNA? Does it have a single chain or a double chain and what is the relationship between the order of its bases and DNA? How are proteins synthesized? Where does the RNA go? Where does it reside? [...] Some of these questions would be easy to answer *just observe!* You would see the order of the chains, the intimate structure of the cell. Unfortunately, the microscope "sees" at a coarser scale. Improve the resolution a hundred times, a thousand times, and so many problems of the biology will become much easier. I'm exaggerating, of course, but biologists would be grateful to physicists and engineers and would certainly prefer that to the criticism that they should use math more.[6]

That was in 1959. Today we have given some answers and we are rewriting the textbooks of biology and medicine. We, microscopists, started to do this with the electron microscope and with those inventions that have made giant steps forward in microscopy. But in our history, I like to point out proudly, that we did it by bringing the optical microscope to the observation of the biological

---

[5] Taken freely from: R.P. Feynman, *op. cit.*

[6] Taken freely from: R.P. Feynman, *ibid.*

nanoscale with the great advantage, using a piece of glass and the "light of the rainbow", that everything can continue to live, to evolve, to change, as we who are observing it.

Since the time of Aristotle (384–322 B.C.) knowledge through the observation of minute things and the use of optics have aroused interest together with the attempt to explain the mechanisms of our own vision. Pliny the Elder (23–79) tried his hand at "observation", in his monumental work *Naturalis historia* describing the cosmos, dealing with geography, anthropology, zoology, botany and medical zoology, mineralogy, metalworking and the history of art. Claudius Ptolemy (100–170) would much later develop a treatise on optics. Over the centuries, a mixture of curiosity and the aspiration to see what the eyes could not see, led to the development of instruments based on the control of visible light using transparent objects such as glass, the use of which in the form of "glassy pastes" dates back to the third millennium BC in Mesopotamia, according to Pliny the Elder. The interest developed among astronomy, medicine and biology and in Italy the results were remarkable. Among the many, we need only think of the work of Girolamo Fracastoro (1478–1553).

The great leap forward was taken thanks to Galileo Galilei (1564–1642) in his elaboration, fundamental to this day, of the scientific method.

It was Galilei and the Dutch who offered the cue to develop not only theories on astronomical observation but also, more importantly, to build an instrument that would allow one to see small things well, close up, just as one could see the sky with a telescope. So, the Galilei's "occhialino" *for seeing the smallest things* becomes the *microscope* in the letter written by Johannes Faber to Federico Cesi on 13 April 1625. In the words of that letter the name,

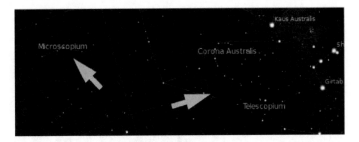

Fig. 2.5 A constellation, *Microscopium*, reinforces the idea that the microscope is a "star" among instruments that study the living. A bit like the telescope for distant stars

forever: "*microscopium nominare libuit*". Its kinship with the telescope could only be a premonition of the fact that that name would also become the name of a constellation (Fig. 2.5). The Microscope, abbreviated as Mic, is one of the modern constellations. It is a small, inconspicuous southern constellation located south of Capricorn and west of the Southern Fish. It was catalogued by Nicolas-Louis de Lacaille, a French astronomer, and made known in a 1763 publication. The optical microscope is therefore a "star" beyond the name of the constellation.

In any case, it was in the second half of the seventeenth century that the optical microscope began to establish itself as an instrument of knowledge. Not by chance: it is the time when the mechanistic conception began to influence the theories on the nature of the physical world, leading to the consideration of all observable phenomena as originating from the mechanical interaction of small bodies that made tiny machines, michromechanisms, mostly invisible to the naked eye. It was therefore, even then as in Feynman's years, indispensable to improve the optical microscope of Galilei and of the growing number of microscopists.

Fig. 2.6 The orderly structure of the monastery where Saint Francis lived in Cortona, "The Cells" hermitage. That order given by the "cells" suggested the name for the "biological cells"

In January 1665, an amazing, unique, monumental book was published. The work is entitled *Micrographia* and contains a collection of images reproducing observations made by the English naturalist philosopher Robert Hooke (1635–1703) with an optical microscope. Hooke's declared ambition was to succeed in providing an empirical basis, therefore founded on the data of immediate experience and practice and alien to scientific rigor, to the mechanistic conception of Nature.

I would like to show you, before going on, a hermitage in the territory of Cortona founded by Saint Francis in 1211, like all Franciscan hermitages immersed in nature and above all in the area were Lorenzo Cherubini, aka Jovanotti, my daughter Claudia's favourite singer, along with Francesco Guccini, lives (Fig. 2.6). Dwell on the picture and on some details. Among the images drawn in

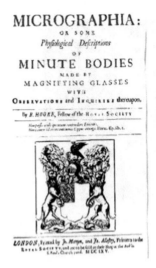

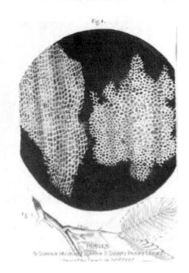

Fig. 2.7 *Micrographia* is the "foundational text". Robert Hooke, a curious physicist who jotted down what he saw under the microscope

Micrographia there is one that, in some way, recalls that monastery (Fig. 2.7).

The sample observed was obtained from cork and it is said that Hooke, in preparing to write the legend, was speechless. It occurred to him, however, that he had just visited a friend who lived in a monastery.

The regular structure of that monastery, made up of many small cells, gave him an idea for writing the figure legend. Thinking back to the cells of the monastery he found a certain similarity with the arrangement of those structures he had reproduced by observing the cork with his microscope. He decided to call them cells. Cells: that's what we call our cells, our biological cells. So, a name that is so popular and fundamental in biology and medicine owes its origin to an observation under an optical microscope.

Fig. 2.8 Embedding a glass ball in a piece of metal to see better, with Antoni van Leeuwenhoek's microscope, as through a dew-drop on a leaf

Between tripods and telescopic lenses, focus adjust-ments and the eyepiece, the optical microscope had several developments. In Italy, Eustachio Divini and Giuseppe Campani were the reference for the construction of micro-scopes and telescopes towards the end of the 17th century. Similar situations occurred in other countries including Hooke's England.

For a long time, the microscope was a derivative of the telescope, until Antoni van Leeuwenhoek, a textile mer-chant from Delft. In the very same country, Holland, of the brothers Hans and Zacharias Jansen, two Dutch spec-tacle makers, that made a microscope which magnified up to thirty times details, inspiring Galilei at the end of the sixteenth century.

However, Van Leeuwenhoek succeeded in creating a microscope with a single lens, set in a piece of metal (Fig. 2.8), which, to see it today, is somewhere between the dewdrop that allows you to better see the veins of a leaf that hosts it and *blips*, the pocket microscope that, by applying a polymer lens to the incorporated camera,

Fig. 2.9 BLIPS takes up the idea of Antoni van Leeuwenhoek today. A polymer lens in front of the photographic eye of the mobile phone transforms it into a microscope (www.smartmicrooptics.com/en/blips/)

makes your mobile phone capable of capturing details up to 800 nm (Fig. 2.9).[7]

In addition to his observations of blood corpuscles and the structure of animal and vegetable tissues, van Leeuwenhoek was responsible for the original description of the lamellar structure of the crystalline lens and the observations of "small animals" in the most diverse liquids

---

[7] M. Cesaretti, A. Diaspro, A. Antonini et al., *Testing feasibility of an accurate microscopic assessment of macrovesicular steatosis in liver allograft biop-sies by smartphone add-on lenses*, "Microscopy Research and Technique," 81(1): 58–63, 2018.

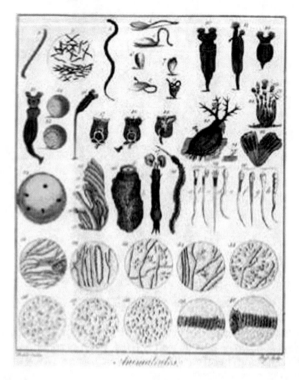

Fig. 2.10 Microorganisms observed by Antoni vandi Leeuwenhoek

(Fig. 2.10). With him, again underlining the close relationship between the optical microscope and the study of living things, protozoology, microbiology and bacteriology were born in a crescendo of scientific rigor and solid experimental bases.

In the nineteenth and twentieth centuries, the number and complexity of questions that were addressed with the use of the microscope grew.

This spreading progressively exposed the limitations of the instrument. Physicists and manufacturers of microscopes were active mainly in trying to offer increasingly

Fig. 2.11 Achromatic microscope Giovanni B. Amici. Photo: courtesy of Sandro Paoletti and Nadia Robotti

high performances in the areas of greatest demand, medicine, biology, biophysics and biotechnology. One of the main aspects of the evolution of the instrument concerned its calibration, that is, the definition of quantitative parameters that link its construction to its use.

Giovanni Battista Amici (1786–1863) was responsible for the direct vision prism and the homogeneous immersion objective, technical innovations that made it possible to correct the chromatic aberration that caused the diffusion of rainbow colours around objects (Fig. 2.11).

A wine merchant, the Londoner Joseph J. Lister (1786–1869), another "vintage microscopist" we might

Fig. 2.12 Abbes law carved in rock at Jena

say, introduced a pair of lenses to correct blurring due to so-called spherical aberrations. These are solutions for optical microscopes and for the cameras you use most frequently.

The real breakthrough, both in terms of quality and in terms of quantitative aspects related to the performance of the lenses and the key parameters for their design and construction, is undoubtedly due to Ernst Abbe (1840–1905). With Abbe, microscopy became truly quantitative in all respects. Abbe developed a true physical theory of the microscope applied with a revolutionary vision by Carl Zeiss (1816–1888) to the design and manufacture of optical components.

Carl Zeiss, the fifth of twelve children of an artisan toymaker, founded the company of the same name on November 17, 1846, at 7 Neugasse in Jena to produce precision optical equipment, particularly microscopes. Right from the start Zeiss focused on product quality and succeeded in establishing itself on the market. In 1866 and 1884 respectively, Ernst Abbe and the chemist Otto Schott became partners in Carl Zeiss, with the idea of firmly basing their activities on the scientific method and product quality. A winning choice and decisive for the development of microscopy and other companies in the sector, from Leitz and Leica in Germany to Nikon in Japan, to name but a few. We owe to Abbe the so-called

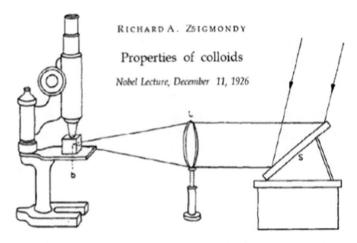

RICHARD A. ZSIGMONDY

Properties of colloids

Nobel Lecture, December 11, 1926

Fig. 2.13 The ultramicroscope, ancestor of the light sheet microscope, was designed to observe the properties of colloids

sine condition (or Abbe's condition) that must be met by a lens or other optical instrument to produce sharp images of objects both off the optical axis and at its center and which, to simplify, says that the sine of the exit angle in an optical system must be proportional to the sine of the entry angle. Abbe, however, became famous for Abbe's formula for optical resolution (Fig. 2.12), the result of an experiment from which he deduced that the ability of a microscope to capture detail depends on the colour of the light, the material through which the light travels and the geometric and constructional characteristics of the piece of curved glass used.[8] From that moment on, an incessant expansion of the optical microscope and its

---

[8] E. Abbe, *Beiträge zur Theorie des Mikroskops und der mikroskopischen Wahrnehmung,* "Archiv für Mikroskopische Anatomie," 9(1):413–468, December 1873.

progressive, continuous improvement began. The light microscope allowed scientific controversies to be resolved, such as the one between Pasteur and Koch on the pathogen of anthrax, it contributed to the 1925 Nobel Prize in Chemistry won by Zsigmondy on the properties of colloids and to the invention in 1903, together with Seidentopf, of the ultramicroscope, a microscope with a lateral view (Fig. 2.13). These are just the "tip of the iceberg" for all that has happened in microscopy in that wonderful relationship between a curved piece of glass and the rainbow.

From Galilei's microscope we will reach the super microscopes, moving in the four dimensions of space and time. We will investigate the living, collecting the messages brought by the visible light, the one under which we live our daily life. Our eyes will see beyond the imaginable and we will be pointillist painters of molecular landscapes, simply using a piece of curved glass to guide the light of the rainbow. Abbe's law indicated in a lapidary way our limits and the Florentine Giuliano Toraldo di Francia (1916–2011) explained us how to transform our sight in super-sight when coined the term "super resolution". From Richard Feynman's lectures, one consideration, of all things, is the picklock for knowledge: "Matter is made of atoms."

We share the six different atoms of oxygen, carbon, hydrogen, nitrogen, calcium and phosphorus for about 99% of our composition. Why are we different? Why don't I look like George Clooney? Because what matters for a stone or a living being, besides the elemental atomic composition of matter, is the organization of atoms into molecules, of molecules into macromolecules, and then into cells, tissues, and organs.

The organization on a very small scale of matter is decisive in the delicate balance of the intricate and dynamic relationship between structure and function. With the black and white keys of the piano we can compose infinite melodies as the sequence of those four pairs of bases, A-T-C-G, is able to realize infinite configurations of the living being that we are. The living being that lives healthy, or that gets sick, and that sometimes heals.

How to understand something about how we are made and how we function? It should be clear by now. "Just observe!" We'll do it like not even Harry Potter could do it, after all it's not Magic but Microscopy.

3

The Colours of the Rainbow

Close your eyes and see All the colours of the dark.
Roberto Vecchioni, I *colori del buio,* in "I colori del buio"
(2011)
There is a crack in everything, that's how the light gets in.
Leonard Cohen, *Anthem,* in "The Future" (1992)

We see in colour what is around us, many years ago we used to watch television in black and white and that was enough, many people imagined colours. When you watch a black and white picture taken by Gianni Berengo Gardin, one of the greatest photographers in the world and honorary doctorate, you do not need colours to be fascinated or to understand what's going on. As well, to the footsteps of the man on the Moon, after all, the colours were not needed (Fig. 3.1). In Italy it was 22:17:40″ on July 20, 1969 when the Apollo 11 module landed on the dusty surface of our satellite. Tito Stagno and Ruggero Orlando made that long black and white night extraordinary. I saw the descent of the first man on the moon in

© Casa Editrice Ulrico Hoepli S.p.A. 2022
A. Diaspro, *Expedition into the Nanoworld,*
https://doi.org/10.1007/978-3-030-94472-8_3

Fig. 3.1 On July 20, 1969, Neil Armstrong and Buzz Aldrin touched down on the lunar surface, leaving a human footprint

Colfosco, Alta Badia, at the Hotel Riposo, huddled on the floor in the TV room together with many other boys and girls whose families had given them permission to exceptionally go to sleep after *Carosello, Italian television advertising show*. Many summers later, Nicoletta Orsomando, an unforgettable italian television presenter, announced on 17 July 1976, on the occasion of the opening ceremony of the XXI edition of the Olympic Games held in Montreal, Canada, that the sports events would be broadcast in colour. I saw those Olympics in colour thanks to Grandpa Mario, who has always been in line with the "mass" technology available on the market.

Now, what about biological cells, which have no colour of their own to show? Is it possible to observe them in colour? Distinguish a healthy cell from a sick one using colours? Or, do we have to imagine the colours to get the hidden secrets of life?

Our lives flow under the sun with the colours of the rainbow. Light governs our days between waking and sleeping. We are bathed in electromagnetic waves and live in that portion of the energy spectrum called the visible.

It seems natural that this radiation has no harmful effects on our health, and so it is. Electromagnetic waves are not, or do not often, get absorbed by the molecules we are made of, otherwise we would be 'cooked'.

Absorption by a material, in our case the biological cell, concerns its ability to capture the energy associated with the electromagnetic radiation that propagates inside. This is the energy of photons, quanta of electromagnetic energy, which is transferred to the electrons, atoms and molecules of the material, transforming it into internal energy, such as thermal energy. Absorption depends on both the nature of the material and the "colour" of the radiation in the case of visible light. The phenomenon of absorption, the absorption of energy such as that carried by electromagnetic waves, would typically produce a rise in temperature and our proteins or DNA would be damaged.

When you make yourself scrambled eggs in a pan, you actually denature the proteins with the heat of the flame, you break the balance between organization of the egg molecules, structure and function. They are tasty, but you have lost the functionality of the protein.

DNA and proteins absorb in the ultraviolet region, between 220 and 280 nm, as do other molecules of life.

This characteristic is relevant for observations and, more specifically, in our case, to observe with the optical microscope what happens to the living being in space and time. In order to see something, we have to realize what is called contrast (Fig. 3.2).

For this reason, when we paint, for example, we use colours, which are nothing more than materials whose molecules absorb all the visible radiation we send out except that which we see by reflection. If you look at the

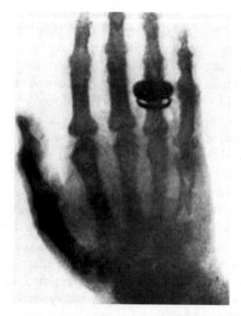

Fig. 3.2 X-ray radiograph obtained by Wilhelm Conrad Röntgen of the hand of his wife, Anna Bertha Röntgen, showing the different contrast of the ring to the tissue and bone. Contrast discovered by Marie Curie, née Skłodowska (1867–1934), the only woman to have won the Nobel Prize twice and in two different areas, physics and chemistry) and the first to teach at the prestigious Sorbonne in Paris. Marie Curie shared the Nobel Prize for Physics in 1903 with Antoine Henri Becquerel and Pierre Curie, and received it alone for Chemistry in 1911

stained-glass window of a cathedral, the light that passes through not only illuminates the environment but makes you see colours and decorations, stories of knights and saints (Fig. 3.3). The reason lies in the fact that those glasses selectively let through the light of the rainbow and its colours, producing a contrast visible to the eye that is sensitive to the square of the modulus of the electromagnetic field, putting together the time it takes for the light to reach the eye and the energy, the colours, filtered by

Fig. 3.3 Stained glass window of the rose window in the Cathedral of San Lorenzo in Genoa

the glass. The speed of light is so high that the eye cannot appreciate modest slowdowns such as those produced by stained glass.

The speed of light is the speed of propagation of the associated electromagnetic wave and is universally denoted by the letter c, from the Latin *celeritas*. In the absence of matter, in the deepest vacuum, between 10^{-9} and 10^{-5} Pascal, light propagates in a straight line at a constant speed of 299,792,458 meters per second which, for simplicity, can be approximated to 300,000,000 m/s or 3×10^8 m/s in free space. To give an idea, atmospheric pressure is about 10^5 Pascal and the "vacuum" produced by the home vacuum cleaner comes to 0.8×10^5 Pascal. There are about fourteen orders of magnitude between the highest vacuum and the one we can imagine, fourteen zeros. The speed of light is a physical constant that is independent of the speed of the object emitting the radiation, is slowed down by matter, and is considered the maximum speed at which any information in the universe can travel. Without considering other characteristics here, we can remember that the speed of light is about 872,000 times that of sound. During a thunderstorm we first see the lightning and then hear the thunder produced by the lightning. Since sound travels at about 330 m/s, it travels 1000 m, one kilometer, every three seconds. So the seconds we count between the lightning and the sound of lightning let us know how far the lightning struck by dividing by three. We will use this tactic later to "decode" space using time.

We return to our problem of obtaining contrast from transparent objects such as biological cells and their constituents.

Bartolomeo Camillo Emilio Golgi (1843–1926), a few weeks before Giosuè Carducci (1835–1907), was the first Italian ever to be awarded the Nobel Prize for medicine in 1906 by the Karolinska Institute in Stockholm.

Golgi's laboratory was a small kitchen in Abbiategrasso where he kept a microscope and a few instruments. He was fortunate to have had as a research teacher and friend Giulio Bizzozero (1846–1901) who, in the best tradition of the University of Pavia, together with Paolo Mantegazza (1831–1910), had founded the histological laboratory to study plant and animal tissues. In 1873, Golgi developed the revolutionary "black reaction" in his kitchen.

He succeeded in selectively colouring nerve cells, and, by perceiving the way they organise themselves.

He received the Nobel Prize ex aequo with Santiago Ramón y Cajal (1852–1934) for his studies on the histology of the nervous system.[1] Golgi's motivation was

---

[1] G. Grant, How the 1906 Nobel Prize in Physiology or Medicine was shared between Golgi and Cajal, "Brain Research Reviews," 55: 490–498, 2007.

the development of the black reaction and Cajal's motivation was the discovery, thanks to Golgi's staining, that neurons are physically separated from each other, i.e. that they interact not by continuity but by contiguity through the synapse. He did not share with Golgi the conclusions about the neuronal network, but together they brought an epic breakthrough in the study of the brain. The key lay in the possibility of observing something hitherto invisible (Fig. 3.4).

When light passes through a transparent object, there is a change in the propagation of light waves. It is called phase change. Our eye, like our camera, is not able to perceive the phase difference.

More simply, it is unable to notice the different speed with which light passes through different, transparent parts of an object as well as a cell.

One eye isn't, but two eyes are. As they say, if you have two unknowns you need two equations to unravel the skein. In some cases, each eye provides one unknown and this, for example, allows us to understand that a person is sitting in the front row compared to someone sitting in the back row because we are able to solve, in our brain, the two equations with respect to the variables of arrival time of the information that accounts for the distance and amplitude of the signal.

Now, since cells are made of material that, although transparent, propagates light at different speeds, the only way to produce a contrast that allows you to build an image is to create a map, point by point, of the different speeds of attraction.

In the early 1930s, Frits Zernike (1888–1966), a Dutch physicist, had the idea that made it possible to record, in the form of light intensity, the different speed of crossing, i.e., the phase variation. A ray of light is passed through the object while another reference ray passes through it.

Fig. 3.4 Neurons brought into contrast using the Golgi technique

And their combination allows to see the invisible by a kind of "differential" contrast. The electromagnetic waves will cancel out or "sum up", creating a contrast that will make it possible to form images of objects, such as the crystal glass, that would otherwise be invisible. A "phase contrast" will be realized.

Zernike was awarded the Nobel Prize in Physics in 1953, for "his demonstration of the phase contrast method, especially for the invention of the phase contrast microscope phase contrast". Again, to allow the invisible to be observed (Fig. 3.5). One can therefore see why the phase contrast microscope has become particularly important in the study of living cells.

How many times did you pay attention to *fluorescence*? Sometimes while observing objects from which, pens or jewels, a coloured light came out, manifesting itself with an attractive brilliance.

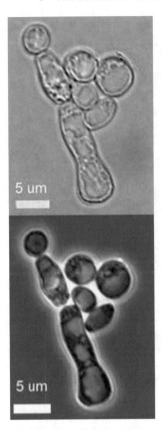

Fig. 3.5 See the in-visible, that which is poorly detailed, top, exploiting phase contrast, bottom. The differences in "phase" translated into intensity changes by Zernike's method

Be amazed at the scenes painted by Michelangelo Buonarroti (1475–1564), a leading figure of the Italian Renaissance, in The *Last Judgement* (1536–1541) at the Sistine Chapel in the Vatican. Among the striking features is the blue of the mantle given by the inimitable blue of the lapis lazuli that adds the brilliance of its own fluorescence (Fig. 3.6). Observing bright colours develops a greater curiosity towards the things that have them.

Fig. 3.6 The fluorescence of the pigment in the colour of the cloak, visible in the lower left-hand panel, makes the Virgin's cloak incredibly bright. "Separata e prossima al figliuolo la madre sua, timorosetta in sembiante e quasi non bene assicurata dell'ira e secreto di Dio, trarsi quando più sotto il figliuolo" (Giorgio Vasari, *Le Vite*, 1568)

Fluorescence, together with phosphorescence, is part of a class of phenomena that go by the name of "luminescence" and when they are characteristic of living beings the term changes to "bioluminescence".

If we look at the terminology in a physical context, we can see that it is a process of light emission and, in general, a process of light absorption electromagnetic waves. Typically, for the situations we are interested in, we will refer to phenomena for which the emission of light is a consequence of energetic transitions of bound electrons which are "excited" in any way, as long as not thermally. This is how we define bioluminescence, the process whereby emitted light has its cause in biochemical processes. Similarly, among the most commonly used nomenclatures, chemiluminescence is associated with chemical reactions, and electroluminescence with electrical phenomena or those produced by electrical charges. The boundary is sometimes blurred. Our main interest goes to photoluminescence, that is the light emission caused by fluorescent and phosphorescent emissions (Fig. 3.7).

Among the phenomena we see most frequently are the light emitted by fireflies, traces of blood made luminous at the scene of a crime, and the glow emanating from coloured star stickers in a dark room. For fireflies a chemical reaction is associated with a biological activity producing bioluminescence. The luminous reaction is generated by the combination of luciferase, an enzyme in the blood, and luciferin, a protein, in combination with oxygen and a natural acid that converts fats and sugars into energy.

Luminol makes traces of blood visible by chemiluminescence. Even if you wash away the blood, the iron in the hemoglobin remains and is enough to trigger a visible reaction with luminol.

Sticky stars are a typical application of phosphorescence. The light charges them with energy, which they slowly release in the form of light.

When you switch off the light you can imagine yourself taking a romantic walk among the stars, or sleeping under the starry sky.

Fluorescence is triggered by light with the peculiarity that between excitation and emission the time scale, which combines, is that of the nanosecond, as can be seen in the diagram in Fig. 3.7. It is necessary to illuminate continuously the fluorescent molecules to obtain continuous light emission. Illumination and emission are distinguished spectrally by the colour emitted according to a precise rule. The second principle of thermodynamics applies and

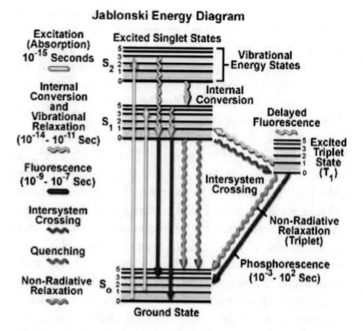

Fig. 3.7 The Perrin-Jablonski diagram represents various types of radiative and non-radiative transitions between energy levels, including intersystem crossing, which fluorescent molecules can give rise to. Olympus website tutorial (https://www.olym-pus-lifescience.com/en/microscope-resource/primer/java/jablonski/jabintro/)

the transformation of light energy is subject to "losses", with emission at a longer wavelength than excitation.

Thinking of the electromagnetic spectrum in terms of energy, if in the visible we go from the "lowest" to the "highest" we find colours ranging from red to blue. A blue illumination will correspond to an emission in the region from green onwards and if we illuminate in the green we expect emissions towards red and so on.

The term fluorescence, derived from fluorite, a naturally fluorescent mineral, was coined by the Irish physicist George Gabriel Stokes (1819–1903) in 1852. The emission of fluorescence by a molecule, generally indicated as fluorophore or fluorochrome, or by a group of molecules is therefore the result of a physical process made of three steps performed in a very rapid succession. Typically, a photon possesses energy E equal to $h\nu$, where h is Planck's constant that is $6.62607004 \times 10^{-34}$ J s and $\nu = c/\lambda$.

This is the energy associated with a certain wavelength, colour, coming from a source of illumination such as a lamp or a laser. It is eventually absorbed, within a certain probability, in a very short time of the order of 10^{-16} seconds bringing the molecule into the excited state (Fig. 3.8).

This excited state has a relatively short time, but you can think of yourself as having received good news, jumping for joy and then allowing yourself to sink into the couch where you were standing when the news broke. What can I say, you have just won the Nobel Prize!

The molecule is in an excited state since it has an electron on a higher energy orbit, this distinguishes it from the excited state of the chemiluminescent molecule which is achieved due to the energy provided by a chemical reaction.

However, the excited state of the fluorescent molecule has a relatively short average lifetime, between 10^{-9} and 10^{-8} seconds. In this dominant time—there are even faster ones—the molecule has several ways to de-excite itself. These include emitting light energy, i.e. photons.

Kind of like if you had a couch, a chair, a hug, or a nice breath to relax.

or a fluorescent molecule, a molecule with the propensity to emit photons after a light stimulus, the excess of energy is dissipated by the release of a photon of lower energy than that of the stimulus, more technically termed excitation. The difference between the energy of the excitation photon and that of the emission photon is one of the

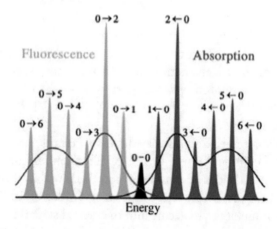

Fig. 3.8 The Franck–Condon and Einstein schemes underlie the fluorescence mechanism. From the quantum mechanical point of view, an electronic transition from one vibrational level to another is more likely the more the wavefunctions corresponding to the two states are superimposed

fundamental properties that characterize a fluorochrome and is referred to as the Stokes *shift*.

From a practical point of view, if you illuminate a fluorescent molecule with a beam of blue light you can select the fluorescence emitted by wearing green glasses. If you happened to uncork a bottle of tonic water under sunlight you could see the beautiful, "celestial" blue fluorescence emitted by quinine and observed by John Herschel (1792–1871) in 1845 (Fig. 3.9).

When the German chemist Adolph von Baeyer (1835–1917) synthesized the molecule Spiro[isobenzofuran-1(3H),9'-[9H]xanthen]-3-one,3',6'-dihydroxy, he described its fluorescence emission as a new class of dye and coined the name "fluorescein".[2] Opening a very wide window on the possibility of studying living things using the phenomenon of fluorescence.

---

[2] A. von Baeyer, Über eine neue klasse von Farbstoffen, "Berichte der Deutschen Chemischen Gesellschaft", 4:555–558, 1871.

A new way to colour transparent objects was catching on. In 1905, Von Baeyer was awarded the Nobel Prize in Chemistry "in recognition of his studies in the advancement of organic chemistry and the chemical industry, through his work on organic dyes and hydroaromatic compounds." Fluorescein, a kind of "magic powder," was a revolution in the dye industry. One of the first uses of fluorescein in 1877 involved a "colossal" and important experiment: the tracing of groundwater in southern Germany that showed unequivocally the connection between the Danube and Rhine rivers through underground streams. Fluorescein, ten kilograms, poured into the Danube reappeared about sixty hours later in a tributary of the Rhine, observed through its coloured emission. The ability to colour the water green is still exploited today in a variation for a fun tradition put on in Chicago, Illinois, for St. Patrick's Day. Hundreds of thousands of people attend the green river spectacle in the "windy city".

It was Albert Coons (1912–1978) who succeeded in penetrating the living by marking antibodies with fluorescein isothiocyanate, giving rise to the field of immunofluorescence.[3] Specific components of the living organism can be highlighted using fluorescence in a biochemically specific way and exploiting the properties of antibodies in terms of biological recognition. In the so-called direct immunofluorescence we use antibodies labeled with fluorescent molecules and, after the time necessary for the specific antigen-antibody interaction, we perform a sort of washing that eliminates what has not bound in order to observe what was of interest for localized emission of fluorescence.

---

[3] A. H. Coons, H. J. Creech, R. N. Jones, Immunological Properties of an Antibody Containing a Fluorescent Group, "Experimental Biology and Medicine," 47, 200–202, 1941, https://doi.org/10.3181/00379727-47-13084p.

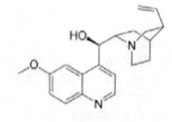

Chinino

Fig. 3.9 Tonic water contains quinine, which emits fluorescence. When you open a bottle of tonic in the summer, under the hot sun, you amaze your friends by noticing the luminescence triggered by solar radiation and the concentration of the substance. Courtesy of Paolo Bianchini, IIT, and Ranieri Bizzarri, UNIPI

The method can be extended to different needs and variations can be used depending on the questions to be answered.

Observing becomes much more interesting if you are able to be quantitative, for example if through the measurement of the intensity of fluorescence you can determine the amount of substance present in the living being. A lot of fluorescence means a lot of DNA, for example.

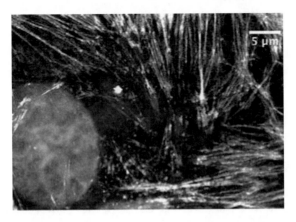

Fig. 3.10 Fluorescence allows specific selection of cellular compartments

Gregorio Weber (1916–1997), an Argentinean scientist and member of the prestigious US Academy of Sciences, enabled the breakthrough of quantitative fluorescent tagging in biological applications.[4] We have at our disposal all the tools to selectively mark cellular components and be able to quantitatively study their distribution and role within individual cells (Fig. 3.10), which will go on to compose organs and tissues of the living. Now brace yourselves because we are going to go through a discovery that will revolutionize the way we can observe living things using visible light and that piece of glass, the lens, which is an optical microscope. We have learned that it is possible to put coloured flags on specific components of the cell to show the arrangement of the various elements. This allows us, with a good knowledge in biology, to obtain indications about the function of that specific organization. The

[4] G. Weber, *Polarization of the fluorescence of labeled protein molecules,* "Discussions of the Faraday Society," 13:33–39, 1953.

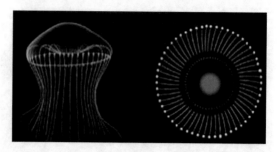

Fig. 3.11 Typical "umbrella" luminescence of *Aequorea victoria*

point is, and was, to insert the flags, the fluorescent molecules, inside the cells and in the right place. From the outside in.

At the end of World War II, in Nagoya, Japan, Professor Yoshimasa Hirata (1915–2000) asked his assistant, Osamu Shimomura (1928–2018), to find out what made the remains of a "shattered" mollusk, Cypridina, glow when moistened in water. Shimomura discovered the reason and proceeded to discover a blue light-emitting chemiluminescent photoprotein, aequorin. The jellyfish *Aequorea victoria* (Fig. 3.11) has a delightful umbrella shape, and the "edges of the umbrella" emit beautiful green light.[5] The jellyfish's agitation triggered a process that would turn out to be dependent on the presence of calcium by producing chemoluminescence.

That blue and green gave one pause to wonder what might be going on in that jellyfish caught on the west coast of North America. A form of light energy transfer from more energetic colours, blue, to less energetic ones, green, was a good candidate. But where was the green

---

[5] O. Shimomura, F. H. Johnson, *Extraction, purification and properties of aequorin, a bioluminescent protein from the luminous hydromedusan, Aequorea* Y. Saiga, "Journal of cellular and comparative physiology," 59:223–39, 1962.

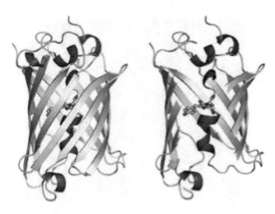

Fig. 3.12 GFP consists of beta-sheets that form a 13-barrel or 13-can structure

bulb? It was, it is, in a barrel-shaped protein (Fig. 3.12) that protects a chromophore, which would become one of the most famous proteins in light microscopy, the *green fluorescent protein*, GFP.[6] In 1988 Martin Chalfie, born in 1947, learned of the existence of a luminous protein during a seminar at Columbia University in New York, where he has his laboratory. In his laboratory he has been carrying out research for years using as a model system *Caenorhabditis elegans* (Fig. 3.13), a phasmid nematode worm, about 1 mm long, which lives in the soil and has phasmids, which are chemoreceptor organs placed on the posterior extremity of the body. It is a model organism much used for the study of the biology of the development and of the apoptosis; about this last aspect, known also as programmed cell death, Sydney Brenner (1927–2019), who received the Nobel in physiology and medicine in 2002 with Robert Horvitz and John Sulston

---

[6] O. Shimomura, *The discovery of aequorin and green fluorescent protein,* "Journal of Microscopy," 217:3–15, 2005.

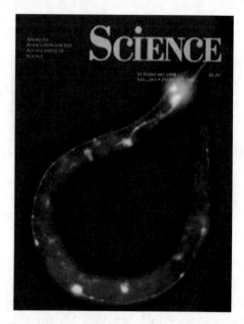

Fig. 3.13 Historic cover of "Science" for the results obtained by Martin Chalfie and colleagues on the utilization of green fluorescent protein

(1942–2018), began experimenting in 1962. From the point of view of light microscopy, the fact that it is a transparent organism makes for interesting observations. Chalfie is interested in the development of the nervous system and has an ingenious intuition. Chalfie's intuition was that by linking the gene for GFP with various specific genetic switches or genes for other proteins, he would be able to observe cells being activated by having them go hand in hand with the gene for GFP. In short, if I look for protein A and I co-express it, I put it arm in arm with GFP, the light signal of GFP tells me where protein A is. The green light becomes a sort of beacon for specific events.

It will be Douglas Prasher, a brilliant scholar who will abandon research to become a used car salesman, who will provide Martin Chalfie with the GFP gene. He started a series of experiments that would result in a fundamental work that would change the way living things are coloured forever.[7] The colour was green, but it didn't take Roger Yonchien Tsien long to figure out how to make a real palette of coloured GFP in his labs in La Jolla, California.

Osamu Shimomura, Martin Chalfie and Roger Tsien will be awarded the Nobel Prize in Chemistry in 2008 "for the discovery and developments related to the green fluorescent protein GFP".

Here, we have a coloured fluorescent protein that can be produced specifically by the living thing we want to study, a class of visible fluorescent proteins.[8]

Kind of like going from black and white television to colour television. Ready to shoot a technicolour film of the living through the organization and function of its proteins: light from the rainbow passes through a piece of glass and illuminates the living, which in turn reveals itself by sending out light signals.

Ladies and gentlemen, "in unified networks" here is the rainbow in the brain produced by a cascade of neurons (Fig. 3.14).

We have "artificial" fluorescent molecules and natural molecules, molecules triggered with light to get light signals: molecules is the key word. Now, in most optical

[7] M. Chalfie et al., *Green fluorescent protein as a marker for gene expression*, "Science," 263:802–805, 1994.

[8] A. Diaspro, *Shine on proteins*, "Microscopy Research and Technique," 69(3):149–151, 2006.

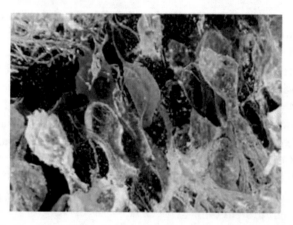

Fig. 3.14 The colours of the rainbow and the neurons in the brain are not an aesthetic fact but indicate the functions and communications taking place

microscopy methods we interact with tens, hundreds of thousands, millions, of molecules at a time. At the tip of our well-hardened pencil we have a kind of average molecule that we bring back into our drawing or with a more or less fine brush on our canvas.

For many years this fact was accepted with the limitation, for example in understanding how a disease develops and progresses, of not being able to access molecular detail. However, this limitation was balanced by the fact that the use of light rather than electron beams allowed for a more widespread method of observation.

It is Albert Einstein (1879–1955) who begins to put a flea in the ear of the scientific community about not only the existence of individual molecules, but especially about their stochastic modes of action, between Brownian motions and pollen considerations.[9] achieving a

---

[9] A. Einstein, *Über die von der molekularkinetischen Theorie Wärme geforder-te Bewegung von in ruhenden Flussigkeiten suspendierten Teilchen*, "Annals of Physics", 17:549–560, 1905.

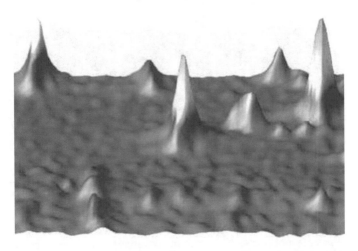

Fig. 3.15 These "peaks" are the signals sent by individual molecules. The ripples represent uncertainty or background noise

fundamental result.[10] At the time, he was working at the IBM research center in San Jose, California. One of the first means of single-molecule detection had been developed in the field of ion channels in biological cells with the development of the *patch clamp* technique by Erwin Neher and Bert Sakmann, both recipients of the 1991 Nobel Prize in Physiology and Medicine "for their discoveries concerning the function of single ion channels in cells." Moerner's experiment inspired many to look at single molecules, among them Eric Betzig who we will meet again later. However, a real leap forward for our microscopy history occurred a year later when Michel Orrit and Jacky Bernard, from their laboratory in Gironde, France, succeeded in showing evidence of the phenomenon of absorption by single molecules through their fluorescence

[10] W. E. Moerner, L. Kador, *Optical detection and spectroscopy of single molecules in a solid*, "Physical Review Letters," 62:2535–2538, 1989.

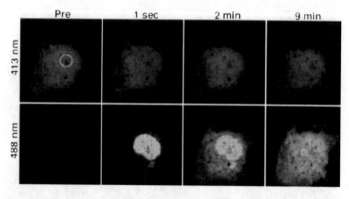

Fig. 3.16 The great insight of photoactivation. As long as a region, visible in the first row, is not photoactivated, its molecules are not visible. The bottom row shows the absence of signal in the first box and the diffusion of "lit" molecules in the subsequent boxes. This ability will be the basis of fluorescence super-resolution and subsequent Nobel recognition

(Fig. 3.15).[11] In 1997 Moer-ner was relocated to the University of California at San Diego where, as it happens, Roger Tsien was trying to make GFP in all the colours of the rainbow. Moerner discovered that the fluorescence of a GFP variant could be turned on and off, on or off, in short it was possible to control its fluorescent behavior at will at the single-molecule level.[12] His experiment was to turn the fluorescent behavior of GFP on and off using light. Betzig's eyes "twinkled" because this would solve his problem of achieving what we call single-molecule microscopy.[13] In the following years, if possible, the story would

[11] M. Orrit, J. Bernard, *Single pentacene molecules detected by fluorescence excitation in a p-terphenyl crystal*, "Physical Review Letters," 65:2716–2719, 1990.

[12] R. M. Dickson et al., On/off blinking and switching behavior of single molecules of green fluorescent protein, "Nature," 388:355–358, 1997.

[13] E. Betzig, *Proposed method for molecular optical imaging*, "Optics Letters," 20:237–239, 1995.

become even more adventurous. Betzig will swing back and forth between research and other occupations until he learns of an amazing achievement by Jennifer Lippincott-Schwartz and George Patterson. They designed and realized a variant, a mutant, of GFP that was initially optically inactive, activated by irradiation at 413 nm, i.e. capable of emitting fluorescence when excited at 488 nm and then terminating its duty, like all fluorescent molecules, by irreversibly inactivating itself due to a phenomenon known as *photobleaching* or fo-bleaching (Fig. 3.16).[14]

In 1989, William E. Moerner became the first scientist in the world to measure the light absorption of a single molecule.

The phenomenon is natural, it also applies to you when you read or listen to someone: after a while you stop doing it, then maybe after a good coffee you go on. Fluorescent molecules allow themselves to be excited many times and many times they emit fluorescence. However, there comes a time when they stop emitting fluorescence for good. It may happen that they stop for a while and then they recover, and this is not true for all of them, but sooner or later they all experience irreversibility. That's *photobleaching*. What turned the tables was the possibility of photoactivating single molecules and controlling the spatiotemporal aspects of light emission.

This ability to control on or off at the level of a single fluorescent molecule, in some cases even on one colour or another, will be the basis of the fluorescence optical super-resolution obtained by Eric Betzig while he was out

---

[14] G. H. Patterson, J. Lippincott-Schwartz, *A photoactivatable GFP for selective photolabeling of proteins and cells*, "Science," 297:1873–1877, 2002.

of work, in the living room of his friend Harald Hess' house.[15]

We'll return to this topic but it's worth anticipating that Eric Betzig will be awarded the Nobel Prize in Chemistry in 2014 along with William E. Moerner. This pair of scientists whose studies have been captured by the fluorescence of single molecules will be joined in the prestigious award also by Stefan W. Hell who, by controlling the phenomenon of fluorescence in its on or off states, will develop fluorescence super-resolution microscopy in a different way. The motivation of the Nobel Committee for Betzig, Hell and Moerner will be "for the development of super-resolved fluorescence microscopy".

The molecules that make it possible to obtain the colours of the rainbow from the living world have opened a veritable new chapter in optical microscopy, suggesting two distinct ways of improving the possibility of capturing ever finer details. On the one hand, there will be instrumental developments of the so-called optical configurations and architectures. On the other hand, developments concerning the control of the probes or preparation methods. The former will always suffer from the limit imposed by diffraction allowing to improve things in a finite way, from three to seven times a finer detail, while the latter will offer and offer unlimited possibilities of improvement allowing us to push the microscope up to spatial resolutions of the order of a few nanometers, 2–4 nm, and precise localizations on the atomic scale, up to 1 Angstrom.[16]

---

[15] E. Betzig et al., *Imaging intracellular fluorescent proteins at nanometer re-solution*, "Science," 313:1642–1645, 2006.

[16] A. Diaspro, *Circumventing the diffraction the limit*, "Il Nuovo Saggiatore", 30 (5–6), 45–51, 2014.

4

A Piece of Curved Glass, the Sharpener of Light

Blind, farsighted, beggars of sight the merchant of light, your ophthalmologist Now wants only special customers Who don't know what to do with normal eyes No longer an optician but a lens dealer To improvise happy eyes So that pupils used to copying Invent worlds to look at.
Fabrizio De Andrè, *Un ottico*, (1971).

Our skin sometimes seems smooth, other times to be cared for and our beard hair grows as we observe it and perform a morning ritual, a caress against the hair on our legs to feel those we do not see. Sometimes we wish we could see better, more, build more detailed images of our surroundings in our minds. From the playground pond we pick up some water in the palm of our hand and we seem to see something moving, are they tadpoles or single-celled organisms, *Paramecium primaurelia* for example? A piece of glass could help us. A well-polished and worked piece of glass. We are not "the man who walks on pieces of glass" of Francesco De Gregori, the "Prince" of the Italian

© Casa Editrice Ulrico Hoepli S.p.A. 2022
A. Diaspro, *Expedition into the Nanoworld*,
https://doi.org/10.1007/978-3-030-94472-8_4

Fig. 4.1 A digital camera allows us to represent the reality we observe by image points, pixels. The idea of the digital photo is courtesy of Paolo Sapuppo, Leica Microsystems

song. We are simply curious. Sometimes a drop of dew is enough.

Let's imagine that we are pointillist painters, and we want to make our painting point by point by reporting on the canvas what we see under visible light (Fig. 4.1).

If the detail we want to report is finer than the capacity of our eyes, we will use a curved piece of glass, a lens. The lens will allow us to see better in the sense that it will allow us to capture details on a scale that, according to the laws of physics, is in the order of 200 nm.

With my eye I can distinguish details of the order of the most technological school ruler, which is down to the tenth of a millimeter, 100,000 nm.

The lens is the main element of the optical microscope, of any optical microscope (Fig. 4.2).

The quality of the glass that is used is important and the choice between *crown* glass, with a low refractive index, or

Fig. 4.2 The beautiful lens allows details to be fixed due to its property of focusing light. This "giant glass ingot" is the symbol of the Nikon Museum in Tokyo, Shinagawa

flint glass, with little dispersion, can be relevant for special applications. The beauty of the lens lies in the fact that by copying nature, from the dewdrop to the shape of our eye, a small, simple curvature of a piece of glass produces an immediate effect on our observation of what is around us.

For our purposes it is convenient, without departing too much from reality, to imagine what surrounds us as made up of many small point sources of light that can be switched on or off.

When light reaches these points, they can (i) absorb it without returning anything in terms of light, they can (ii) absorb it and return luminescence, fluorescence, or they can (iii) reflect or diffuse it.

We are there, and we capture the response of matter to light to put it on our canvas using a coloured pencil, a brush or something to light up an element of our digital screen, a pixel, the elementary unit of the digital image. We will put the dots in the order of the observed scene to get the final painting (Fig. 4.3).

Fig. 4.3 The pointillist painter captures point by point the details to be put on the canvas and the fineness of the detail depends on the width of the hair and the shape of the point. Georges-Pierre Seurat, *Un dimanche après-midi à l'Île de la Grande Jatte* (1883–85), oil on canvas, 207.6 × 308 cm, The Art Institute, Chicago

We are now interested in understanding how a piece of curved glass can help us. We have still to make a little effort of imagination. For every point we want to represent on our canvas we must consider that the lens allows us to capture the detail in its own way.

A bright dot after the lens is no longer a dot. It resembles the ripples in the water when you throw a stone into a pond, or the few ripples produced by Tania Cagnotto in one of her perfect dives from the springboard (Fig. 4.4). Tania Cagnotto the first female Italian diver to win a medal in a World Championship in 2009 and Olympics medals.

Now, when the point like emitters we want to represent are too close together, it creates a kind of image of confusion. What is important is not so much the dimension

Fig. 4.4 An "almost" perfect dive by Tania Cagnotto, almost "punctiform", with few ripples

of what we want to represent but the distance of separation between the objects that allows us to distinguish them (Fig. 4.5).

The shape assigned to the detail we want to represent is characteristic of the properties of a certain lens. Since in order to receive a light signal, we must also send light, this

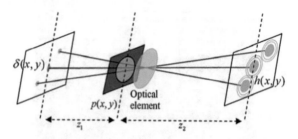

Fig. 4.5 The proximity of point emitters can produce a confusion image

characteristic applies both to the precision with which we illuminate our target and to the way in which we collect its signals (Fig. 4.6).

If you think of the stars in the sky, of a starry sky on a clear summer night, due to the diffraction that results from a uniformly illuminated circular aperture, you can see the stars in the sky.

If you look carefully, you can see a bright region in the centre that, together with a series of concentric rings, is called Airy's disk (Fig. 4.7), in honor of Sir George Biddell Airy (18011892), English royal astronomer between William IV and Queen Victoria of the House of Hanover, between the first and the second half of the 19th century. The characteristics of Airy's disk are a function of the colour of the light and the diameter of the circular aperture through which the light passes. Airy's disk is important both in optics and astronomy. Because of diffraction, the smallest point at which a ray of light can be focused using a lens is the size of the Airy disk to which even the most perfect lens must be subjected. Considering the light passing through an aperture of diameter a, a figure is formed made up of light and dark regions that can be detected, for example, on a screen placed at a certain distance from the aperture, as we have just seen. Away from the aperture,

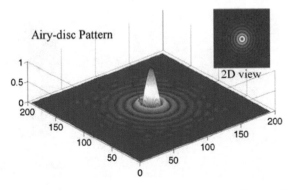

Fig. 4.6 "Diffraction" applies in the same way by illuminating or detecting light

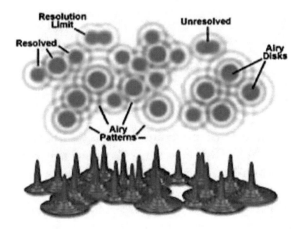

Fig. 4.7 Light intensity distribution of an Airy "figure", which applies to distant stars or individual photon-emitting proteins

the angle at which the first minimum, the dark part, is "seen" along the direction from which the light comes is obtainable from the relation $\sin\theta = 1.22\ \lambda/a$, where λ is the wavelength of light and a is the diameter of the aperture. Rayleigh's criterion says that to succeed in resolving two objects, the center of the Airy's disk for the first object

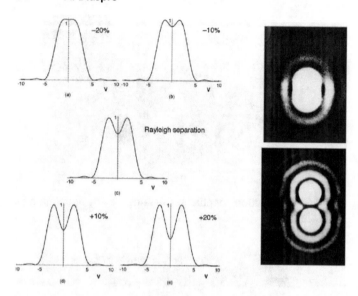

Fig. 4.8 Considering the Rayleigh criterion, when the separation between point sources is lower (−20% or −10%) or higher (+20% or +10%) the distinctiveness decreases or increases

must be in the first minimum of the Airy's disk for the second (Fig. 4.8). This determines the angular resolution of a diffraction-limited system and consequently the ability to distinguish in space objects separated by a certain distance. For this reason, when designing and building an optical instrument we try to limit as much as possible the size of the central disc of the diffraction figure. It is precisely the Rayleigh criterion, the angular relation of "light-dark", that suggests that to make the optical instrument more precise it is necessary to increase the diameter of the lens or decrease the wavelength going, in our case, towards blue. Telescopes with very large lenses or microscopes a blue light therefore offers the best performance, remaining in the visible region of the electromagnetic spectrum.

Wanting to consider directly the case related to the optical microscope, we will treat the piece of glass, the lens, in the same way concerning the two directions in which light travels. This applies when the lens illuminates an object or when it collects the light. To avoid rigorous theoretical treatises, which are exciting but outside the scope of this story, let us try to simplify the scenario with some operational considerations that can be generalised.

Therefore, the scene to be observed is formed by point sources that receive illumination and, in turn, emit light. We would use the so-called Abbe's law, even if Abbe did not explicitly formulate it in the form we use it.[1,2]

Abbe's law defines what is the minimum distance d within which, in the scene we are observing, two objects can be considered as distinct. Abbe's law, carved in stone, states that $d = \lambda/(2n \ldots \sin\alpha)$. This agrees with Rayleigh's criterion and the considerations made about Airy's disk. This distance that we call d tells us how fine is the tip of the brush with which we are going to make our pointillist painting. The lens behaves like a pencil sharpener for light, it shapes it by refining its form (Fig. 4.9). So, d is the minimum distance within which we can say that we are in front of two distinct objects and we refer to this as the spatial resolution of our piece of glass, of the lens, of the microscope as a whole.

So, we now know that the spatial resolution of our optical microscope is directly proportional to the wavelength of light. Colours towards red offer a lower chance of capturing details than colours towards blue.

[1] C.J.R. Sheppard, *Resolution and super-resolution*, "Microscopy Research and Technique," 80:590–598, 2017.

[2] E. Abbe, *Beiträge zur Theori des Mikroskops und der mikroskopischen Wahrnehmung*, "Archiv fur mikroskopische Anatomy," 9:413–418, 1873.

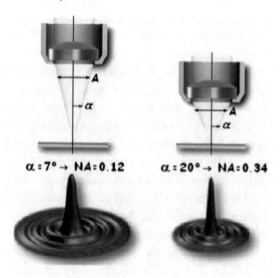

Fig. 4.9 The lens behaves as a "sharpener" for light, making it more or less "sharp" depending on the geometric character-istics and molecular properties of the matter passed through, with the same colour and wavelength used. Freely taken from the Nikon educational site, https://www.microscopyu.com/micro-scopy-basics/numerical-aperture

However, red colours offer the possibility of penetrating more into matter, into the living. If you have two "laser" lights, like the ones you can buy for cheap at any market, one green and one red, and try to pass the light from the fingertip to the nail, you will notice that the green light spreads less. This is because light, as it passes through matter and its constituents, is diffused according to a law that goes, in first approximation, with the inverse of the fourth power of the wavelength. Red light scatters much less than green light so it reaches targets at greater depth, all other things being equal.

Returning to *d*, we note that the spatial resolution is inversely proportional to two interesting quantities, namely if their values increase, the brush tip becomes finer, as we have seen in Fig. 4.9.

The magnitude $\sin\alpha$ is related to the geometry of the piece of glass, to its curvature. Imagine many beams of light all lined up and arriving at the lens, and the lens directs them towards a single point called the focus of the lens. If we want to be rigorous, things are not so simple, but they give a good idea both from an intuitive and a practical point of view (Fig. 4.9). The angle that is formed at the vertex of the triangle which represents the convergence of the light beams starting from the point where they meet the lens, we consider half of it and we call it the half-angle α' of the lens opening. The spatial resolution is therefore inversely proportional to the value of a trigonometric function limited to the unit value. So, no matter how large the angle is at the maximum, the situation does not improve. It can only get worse since $\sin\alpha' < 1$, and in any case α' equal to 90° would mean dealing with an infinitely wide lens.

Then we notice that in Abbe's formula matter also enters, that matter made of atoms that can almost crowd the space between the piece of glass and the scene we want to observe. Matter is represented in the formula by the letter n, the index of refraction of the medium that defines its molecular properties and its influence on the variation of the speed of light. The higher the refractive index, the "denser" the matter and the "slower" its crossing. Therefore the "vacuum", which is not "crowded", is the reference with a value equal to 1. Light does not slow down in a vacuum; its speed is c. A molecule is a set of two or more atoms connected to each other through a chemical bond. The way the atoms organize themselves defines the properties of the molecule, hence the functions, and the combination of the molecules defines the material with its characteristics. Remember Richard Feynman's message that "matter is made of atoms" (R.P. Feynman, *Six Easy Pieces*).

The refractive index *n* of a material, of a medium, is a dimensionless quantity that accounts for the variation of the propagation velocity of electromagnetic radiation when it passes through a material, with more or less dense or crowded sets of atoms. We can evaluate the variation of its direction, according to the phenomenon of refraction. The refractive index is the ratio between *c*, the so-called speed of light in vacuum, which is 299,792,458 m/s, and *v*, the speed of light the medium: $n = c/v$.

When we deal with optical microscopy, we refer to a condition in which matter is transparent to rainbow light rays and is not too dense. We treat matter with a refractive index of slightly more than 1 and less than 2. The value is dimensionless since it is a ratio between velocities.

The index of refraction varies with the variation of the wavelength of light, its colour and, more generally, the frequency of oscillation of the electromagnetic field that passes through the material. We will deal mainly, almost exclusively, with the biological cell, the molecules that compose it or the tissues and organs that the cells form. We can refer to average values of *n*, valid with good approximation in the interval covered by light in the visible portion of the electromagnetic spectrum.

Water, the major constituent of the biological cell, has a refractive index of 1.33, glass between 1.52 and 1.59, oil close to 1.52, air under normal conditions can be approximated to 1, ice to 1.31 and diamond to 2.4. These values are valid, precisely, when the light is yellow, that is, its wavelength is about 590 nm.

Then, considering Abbe's law in its classical formulation, $d = \lambda/(2n \ldots \sin\alpha)$, we note that if we fill the space between the lens and the object that we want to observe with a homogeneous medium other than air the situation can improve, that is, d can decrease. We can use water or oil realizing, respectively, a homogeneity of material along the optical path with the cell or with the glass of the lens.

Typically, in the visible and for very good lenses, the value of d can be considered around 200–250 nm. You can practice this by considering green light (570 nm), water ($n = 1.33$) and an aperture half-angle α of about 60°. The use of a lens improves by about five to five hundred times the ability of our eye to distinguish details to the tenth of a millimeter, allowing us to observe the cell in its internal components. So, we have another piece at our disposal for observing the living, the lens. Our light sharpener.

As you can understand from Abbe's law, to improve d *we do* not have many weapons at our disposal. We can "shorten" the wavelength by pushing the limits of the rainbow on the side of the blue, towards the violet, avoiding the ultraviolet so as not to damage the living being during the observation. If we try to increase n *we realise* that the matter, the medium between the lens and the sample to be observed, becomes denser as n increases and if n were to exceed, for example, the value of 2 we could find ourselves in conditions of turbidity as well as excessive "deviation" of the light rays. The effect may be interesting for a jewel but not necessarily for an optical instrument used to observe the living. So, beyond $n = 1.5$, typically, we do not go.

On the other hand, values of 1.5 or 1.3 are "good" values because they deflect less light beams by being close to the value of the lens glass and the biological cell.

It is worthwhile, now, to dwell on the role of the medium between the lens and the specimen, the passage of light between the piece of glass and what we want to observe.

If everything were homogeneous, a homogeneous index of refraction against a homogeneous organization

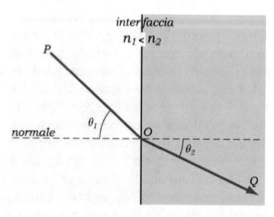

Fig. 4.10 An observer placed in P sees the light ray deflected as in Q, similarly if he puts an outstretched arm in a tub of water. The refractive index of water is about 1.33 versus about 1 for air

of matter, the rays of light emerging from or entering the lens would not suffer deviations due to the medium or the sample.

The deviations are due to the phenomenon of refraction that can be treated in a relatively simple way using the instrument, more than sufficient for us, of geometric optics. The macroscopic observation that can help you discern what happens, or seems to happen, when you immerse a rod in a tub full of water. You will seem to see the rod bend. When you pull it out of the tub, it is apparent that it has not bent at all. This is an optical effect, like that which occurs when light passes through different media in our microscope (Fig. 4.10).

For the phenomenon of refraction we refer, without making the matter too complex, to Snell-Descartes' law and to the principle of Fermat. Fermat's principle, also known as the "principle of least time", says that of all the possible paths that a ray of light can take when it goes from one point in space to another, it takes the one that takes the shortest time. Doesn't that sound reasonable?

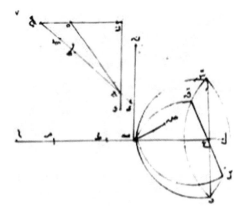

Fig. 4.11 Page from Ibn Sahl's manuscript illustrating his discovery of the law of refraction (circa 984)

While the principle of Huygens-Fresnel is useful to explain the diffraction associated to the deviation of the trajectory of light treated in terms of propagation of waves when they meet an obstacle in their path, Fermat's principle can be used to describe the properties of light rays passing through different media and possible reflections reaching an interesting synthesis in Snell-Descartes' law. This law actually appears for the first time in a manuscript written around 984 by the Persian mathematician, physicist and optician Ibn Sahl (Fig. 4.11). Ibn Sahl (940–1000) tackled the problem in order to obtain the profiles of aspherical lenses, i.e., lenses that focus light by reducing geometric aberrations. Ibn Sahl had access to an important treatise, which he become aware of by consulting Tolomeo's *Optics* (127). In 1621 Willebrord Snel van Royen (1580–1626), known as Snell, reworked the study of the propagation of light rays in different media without publishing it, and one had to wait for René Descartes (1596–1650), for its final elaboration (*Discours de la méthode*). This sets an interesting milestone about the phenomenon of light

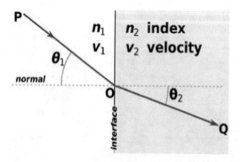

Fig. 4.12 We know that $v = c/n$, the speed of light in a medium of refractive index n decreases if $n_2 > n_1$. This "forces" it to a variation of trajectory respecting Fermat's principle that of all possible trajectories light uses the one that takes the shortest time

passing at the interface between two different media, i.e., with different refractive indexes. Figure 4.12 shows two media with refractive index n_1 and n_2 in contact with each other through a surface, which is called an interface, represented by a vertical line. In the case $n_2 > n_1$, light has a lower phase velocity in the second medium and Fermat's principle must be respected. Snell's law gives the relation between angles θ_1 and θ_2, $n_1 \sin(\theta_1) = n_2 \sin(\theta_2)$. Although the path of the light ray in the two mediums appears to be interrupted, it is as fast as possible given the different refractive indices. In the calculation of the travel time, Fermat said that we must consider the fact that the speed of propagation of light in an optical medium is equal to the speed of light in vacuum divided by the refractive index of the medium. From here we can analyze different situations and consider various approximations. We do not consider, for example, that in the case of non-homogeneous media the light rays can be divided into several parts, as happens in crystals because of the so-called *birefringence*. Now, this is what we need to consider about the effect of a piece of glass between us and the reality to be explored.

However, it is worth considering a phenomenon that Sir Isaac Newton (1642–1726) had brilliantly addressed in his treatise on *Optics*. The phenomenon is that of total reflection, but we do not intend to deal with what happens in front of a mirror, it would make us deviate too much from our histories of optical microscopy.

When light is passed from a denser medium to a less dense one, consider $n_1 > n_2$, it can be verified that the Snell equation is solutionless when θ_1 exceeds a value that is called critical angle, $\theta_{crit} = \arcsin (n_2/n_1)$. In this case, when $\theta_1 > \theta_{crit}$, no refracted ray appears since the incident light on beams a total reflection near the separation region, at the interface. If this happened as it appears to the eye, the discussion would have to deal with what is called the discontinuity of the solution. There is continuity, experimentally verified. A surface wave is generated, called an evanescent wave, which decays exponentially within a short distance from the interface within the medium with refractive index n_2. A sort of skin effect of film made by the electromagnetic field that carries light. The "film" is referred to more formally as the near-field region. At the boundary between glass and air we would find ourselves in the condition of total reflection that Newton had already described when he wrote that "glass attracts the rays of light that try to exit at the Boundary", at the glass-air interface. However, if you were to place a glass sphere of equal "attraction capacity of glass for light rays", you would have the passage of light at the point of contact made in the region of the near field. Newton did not have the mathematical tools to describe the phenomenon but enough genius to understand its scope and mode.

In short, when we send light onto an object to be observed and we set up our light sharpener made from a piece of glass, a series of phenomena are triggered that allow us to focus the light on a point and to read the luminous information from that point, the luminous messages. The dimensions of that point are defined by the laws of physics through Abbe's formulation and so is the extent of the region from which we collect the information. We could go into a precise analysis of what happens using appropriate models and mathematical tools,[3] but we prefer to use a more intuitive approach while maintaining the rigor required by the treatment.

So, let's put on the shoes of the pointillist painter and equip ourselves with a fine brush and a palette of colours. Let us observe reality and try to bring it to our canvas. Our eyes will use the lens of a microscope to better see the detail, to better distinguish the messages sent to us by the many point sources we have to deal with. Do you find similarity in Albert Einstein's portrait (Fig. 4.13) in that irreverent pose of his that has become iconic? Is it faithful? What is the relationship between the "real" Einstein in the photograph and the one depicted in the painting? Microscopy as a song, in the words of Francesco Guccini is a kind of "gibigianna", a flash of light reflected on a surface by a mirror, glass or liquid. A flash of light reflected "on a surface that does not stand still, it expands, travels, conquers territories and people, with plots that escape, through landscapes that you ignore... Because song, pardon! microscopy is magic, a continually migrating phenomenon" (Francesco Guccini, *Note di viaggio*).

Here we are now dealing with the problem of how to trace the relationship between what we observe through the lens and what the observed is really. We call the real image the reality and the observed image what we perceive through the lens. For now, we are not concerned with the

---

[3] M. Born, E. Wolf, *Principles of Optics: Electromagnetic Theory of Propaga-tion, Interference and Diffraction of Light*, Cambridge University Press, 1999.

Fig. 4.13 A mosaic and curious picture, where in spite of the irritating pose you are able to recognize Albert Einstein

so-called contrast mechanism chosen. Whether it's a portrait in charcoal or in tempera colours, on canvas or on cardboard, it doesn't matter. We don't worry about what kind of interaction with the rainbow produces the messages useful in forming the image, but what are the effects of the piece of glass in allowing us to observe the real. The mathematics that describes what does it happen and what we will need for our story that can be read as the text of a story, developing, hopefully, insights at the end of the reading. A bit of mathematics then!

We use the function $i(x, y, z, t)$ to describe the characteristics of the object, the cell for example, that we are interested in observing in greater detail. The choice of the letter i reminds us that this is the "true image" or, for the circumstances that we are going to treat, an "input" signal for our microscope, a set of glass pieces. The coordinates

(x, y, z) refer to the fact that the space we refer to involves the three spatial dimensions organized as a Cartesian term along the three orthogonal axes. The coordinate (t) emphasizes that we are interested in a fourth dimension which is time, which we will deal with later. For all the functions we will mention the first time, we will consider that we are intrigued by whatever happens in the four dimensions (x, y, z, t). At some point in our discussion we will use only one generic coordinate (ξ) in order to re-refer to (x, y, z, t) and not burden the writing. When necessary, we will clearly say which specific coordinates we are referring to. Similarly, the function $o(x, y, z, t)$ will describe the "observed" signal, an "output" signal.

Suppose now that our microscope can be represented with a "box", the microscope model, which has an input signal and an output signal (Fig. 4.14).

So, it seems reasonable, in our case, to consider as input signal $i(\xi)$ and as output signal $o(\xi)$.

What is the relationship between $o(\xi)$ and $i(\xi)$? For our eyes this equivalent to saying "how much and how does the image observed through the piece of glass faithfully represent the true image?".

The optical microscope is responsible for the relationship between $o(\xi)$ and $i(\xi)$. We describe this situation by means of a function $h(\xi)$, which basically represents the effect a microscope has between our observation and the real world. We are interested, now, only in the spatial coordinates (x, y, z). The function $h(\xi)$ describes, then, the way in which the microscope "distorts" reality by, for example, diffraction.

An optical microscope can be treated as a "linear space invariant system". We are interested in two aspects of this definition.

The fact that the system is linear greatly simplifies the treatment, which is necessary at least in broad terms. This

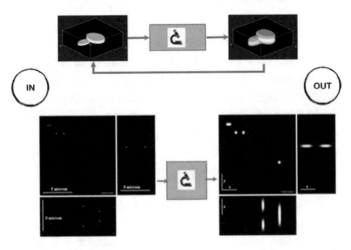

Fig. 4.14 The model relating to the formation of an image is, at bottom, simple. The "real world" enters the microscope to be studied in greater detail and is offered to the observer as a mediation between the "real" and the "observed". Knowing the way in which the microscope modifies reality, it is possible to arrive at the best possible estimate of reality

fact allows to use the principle of the so-called superposition of effects. In mathematics and physics, the superposition principle states that for a linear dynamic system the effect of a sum of perturbations, or input signals, is equal to the sum of the effects produced by each single perturbation, or signal. If you consider reality, as we have already said, formed by a set of point sources that send the light signals that we use to build our pointillist painting after they have passed through the lens, you can think of making your painting by linearly summing the contributions of each single point of the image. Moreover, the hypothesis, reasonable at first approximation, that the system is also invariant in space allows us, once we know the distortion suffered by a point due to the effect of the lens, to measure or theorise, referring to a single point, what happens to all the points of our image.

The function $h(\xi)$ represents just that, what happens to a single point source in crossing the lens. To know $h(\xi)$ we can measure or theorise it; it is enough to send to the microscope or to consider at the input of the "box" that represents the microscope a signal that represents an infinitesimally small point to which we assign, for example, the unitary value only in that precise point and the zero value elsewhere. The function that describes that point signal is called the Dirac function, or impulse, and is indicated with $\delta(\xi)$.

For the properties of linearity and invariance that we mentioned before, if we put a Dirac impulse at the system's input, what we measure at the output will be just the function $h(\xi)$, the distortion that each point undergoes due to the effect of the piece of glass between us and reality.

Noting this feature that can be measured, for example trying to observe a luminous sphere much smaller than the limit imposed by diffraction, so with a diameter of 50–100 nm, or simulate, $\delta(\xi) = 1$ at a certain point, the point where the point source is, and zero elsewhere, we can understand how to write the relationship that binds $i(\xi)$ to $o(\xi)$.

Mathematically, we need to introduce the concept of convolution, which requires the use of integral calculus, out of the scope of our discussion. However we can say that, given two functions $f(t)$ and $g(t)$, the convolution between them is the result $r(t)$ of the operation $r(t) = (f^{\circ}g)\,(t) = \int f(t-x)g(x)\mathrm{d}x$. To simplify the writing, we use the symbol $^{\circ}$ called the convolution operator, instead of the integral relation that sometimes frightens even if it is related to the calculation of a simple area we call $(f7g)$ the convolution product. You can simply try to make a convolution by drawing a function on a sheet of paper and in sliding it over the transparent outline of a second

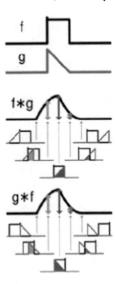

Fig. 4.15 The convolution between the two functions *f* and *g* can be represented as a shift in the chosen coordinate that produces point by point a signal equal to the area shared by the two functions

function just take note of the area in common between the two (Fig. 4.15). This is the operation that the lens does to put together the input signals through the microscope and allow us to observe the image.

If one of the two functions was a Dirac function you would have the other function identical to itself without distortion. Unfortunately, the function representing the lens is a "belly" Dirac with all that that entails.

In our case, where the microscope is treatable as a linear system and invariant in space, we can write that $o(p) = (i7h)(\xi)$. Since, as an effect during the passage of signals through the lens, there could be some "noise", for example someone turning on a light, we make the assumption that we can deal with an unwanted signal that we call n and that it can be summed in first approximation. We have then answered the question "what is the relation between

$o(\xi)$ and $i(\xi)$?". So? Answering this question allows us, known $o(\xi)$ and $h(\xi)$, to go back with an inverse calculation procedure to the best estimate of what we are interested in, that is the best estimate of $i(\xi)$, the true image, having added a det-cut, thanks to the lens, that our eyes did not see.

With little imagination, the operation that leads to the best estimate of $i(\xi)$ is called deconvolution. Since we consider noise to be additive, it is simply eliminable by subtraction once it is determined. How to determine it? Turn off for an instant all the point sources of the image and you will have $o(\xi) = n$.

Since $h(\xi)$ represents the impulse response of the system, you can imagine it as the trace already left by the stone thrown in the pond or the ripples of the water at the entrance of Tania Cagnotto; she, the impulse and the ripples $h(\xi)$. The function $h(\xi)$ is often known as a PSF, the point *spread* function, in Genovese we would say "spatasciata" or "squished", in short, a fried egg gives the idea.

I ask you for a little more "mathematical" effort to describe what the optical microscope does. After this step, you will see, it will be all downhill.

Since every phenomenon that exists in nature can be described through a harmonic analysis,[4] we enter a new mathematical elaboration. We rewrite the mathematical relations we have just seen in what is called the "domain of spatial frequencies" or by applying a mathematical transformation known as the Fourier transform:

$$\hat{f}(\zeta) = \int\limits_{-\infty}^{-\infty} f(x)e^{-2\pi ix\zeta}dx$$

---

[4] R. Bracewell, op. cit.

The Fourier transform describes the trend of a function through a decomposition into harmonics. This formalism, although it seems "difficult", simplifies our discussion. The functions to which we apply the Fourier transform are rewritten with a capital letter, while the generic coordinate ξ becomes a spatial frequency $w\xi$. Recalling a scholastic reminiscence like trigonometry, I think we all remember the sine and cosine functions with their soft, sinuous course. Imagine to be able to describe a certain function as a sum of sinusoidal functions, for example, giving a different weight to oscillations of period or different frequency (Fig. 4.16). As far as our "immediate" interest is concerned, we will only notice that the space variable becomes a space frequency. So, if we call F{} the Fourier transform operator, we can write, without going into the mathematical rigour that can be found in any specialized textbook,[5] that $O(w\xi) = F\{o(\xi)\}$. If we apply the Fourier transform to the relations just discussed and exploit the mathematical property that the Fourier transform of the convolution of two functions is equal to the product of the Fourier transforms of the two functions, we obtain that $O(w\xi) = I(w\xi)-H(w\xi)+N$.

Since the Fourier transform of the $O(w\xi)$ is given by the product of the transform of the *PSF*, $h(\xi)$, for the transform of the $i(\xi)$, we can analyze and study the way in which the individual frequencies are "transmitted by the system" microscope in terms of the so-called MTF, *modulation transfer function*, or the relationship between the modulation of the image and that introduced by the microscope. The *MTF* is nothing more than the Fourier transform of the *PSF*. If you find it complicated think

---

[5] R. Bracewell, *ibid*; J.W. Goodman, *op. cit.*

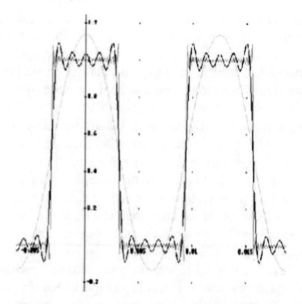

Fig. 4.16 There are an infinite number of harmonics into which you can decompose a function. If you have a square wave, as in the figure, the more harmonics you use to describe it, the more faithful the description will be. If you use only one harmonic, you will have a square wave approximated by the sine function, which in some cases may be sufficient

about music. Think about what you do with your home hi-fi stereo when you're trying to adjust the bass, treble and intermediate sound frequencies.

Think of the feature that most nitpickers ask for when buying an amplifier. It's the MTF, the way in which the frequencies of the sound produced by musicians and recorded on your vinyl, CD or Blu-ray are presented to your ear; if v is the frequency of the sound that propagates in time rather than space then your stereo will have this characteristic $MTF(v) = s(v)/m(v)$. So, the MTF represents the ratio between the sound in terms of frequency $s(v)$, which arrives through the stereo and the frequencies, $m(v)$, produced by the musicians.

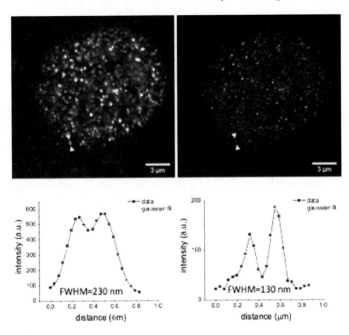

Fig. 4.17 Using our microscope model, having recorded an image like the one on the left and having evaluated the minimum distance within which two points are recognisable, through an inverse procedure we can go back to the best estimate of the "real world" that generated it, the image on the right, where the detail is therefore finer

If we return to our microscope, to get the best estimate of the true signal in spatial frequency contained in the object we want to observe, I, through the knowledge or measurement of H and the recording of the observed image, O, it seems sufficient a simple division, $I = O/H$, leaving out for convenience N. We will see that it is not so simple, but Fig. 4.17 shows us an example of what you can do with the calculation, solving what is called the inverse problem.[6]

---

[6] M. Bertero, P. Boccacci, *Introduction to Inverse Problems in Imaging*, CRC press, 1998.

5

A World in Three Dimensions

I paint landscapes, and I paint nudes
I contain multitudes.
Bob Dylan
(I contain multitudes, Columbia, 2020)

Flatland, by Edwin Abbott (*Flatland: A Romance of Many Dimensions*, 1884), is a multi-dimensional fantasy tale about the life of an inhabitant of a possible two-dimensional universe who comes into contact with the inhabitant of a three-dimensional universe. The encounter between a square and a sphere is compelling. When we make our pointillist painting, even in the finest detail, we make the three-dimensional world in which we live and move a square or a rectangle. If we try to grasp the third dimension, we can't always tell who is above and who is below, who is near and who is far. How can we understand when a tumor cluster is getting big, how many cells it is made of and how they communicate with each other? Our cells are not standing alone, let alone lying on

© Casa Editrice Ulrico Hoepli S.p.A. 2022
A. Diaspro, *Expedition into the Nanoworld,*
https://doi.org/10.1007/978-3-030-94472-8_5

a piece of glass. They're nanomachines, made up of DNA and proteins, continuously processes information concerning so-called mutual interactions. The study of cellular aggregates with the prospect of doing so directly on the "patient" not only reminds us that the microscope is unique in being able to study the details of living things over time, but also that visible light is able to penetrate tissues and organs and study cellular aggregates.[1]

We can therefore explore the world in the three spatial dimensions (Fig. 5.1), let's see how.

Let's make it clear that when in microscopy we use the term sample we often refer to a cell or a set of cells that has a "thickness" of at least 50 μm. The single cell, the one typically deposited on the microscope slide, has a thickness of a few μm. If you try to imagine filling this space with point sources of a few nanometers, or even a few tens of nanometers, you immediately realize that you are in the middle of the party given by Holly Golightly—Audrey Hepburn in *Breakfast at Tiffany's* (1961)-in the apartment at 169 East 71st Street in New York City (Fig. 5.2). A thick sample, rich in fluorescent molecules is like a lot of light given by the emission of many point sources that become almost indistinguishable both in the observation plane and in the whole occupied three-dimensional space (Fig. 5.3).

Let us consider the situation in which we illuminate our three-dimensional sample by sending light through a lens in a certain plane of focus.[2] The light enters the lens and forms, in space, a sort of hourglass with its apex at the point of the geometric focus of the lens.

---

[1] Quoted in J.D. Barrow, *The Numbers of the Universe*, Mondadori, 2003.

[2] B. Bianco, A. Diaspro, *Analysis of three-dimensional cell imaging obtained with optical microscopy techniques based on defocusing*, "Cell Biophysics", 15(3):189–200, 1989.

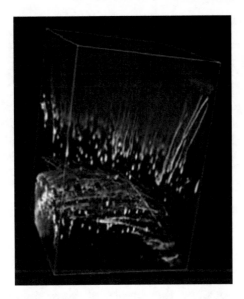

Fig. 5.1 Fluorescence image of neurons in a thickness of about 2 mm with submicrometric resolution. *Credits* Takahiro Deguchi, Ryu Nakamu-ra, Paolo Bianchini—Diaspro Lab and Nikon, Japan

Fig. 5.2 The talented and beautiful British actress Audrey Hepburn (1929–1993) in *Breakfast at Tiffany's* (Blake Edwards, 1961) moving through a crowded environment

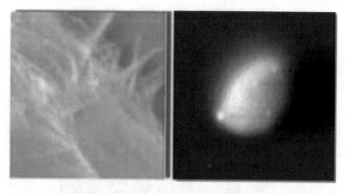

Fig. 5.3 Fluorescence images made fuzzy by molecular crowding in three-dimensional space

Now, if the wavelength used has an appropriate energy to excite the fluorescent molecules contained in the volume occupied by the sample, all those intercepted by the light beam, in any point of space in focus or out of focus, will emit light. Let us imagine that we can "slice" with light a certain number k of different planes in our three-dimensional object and let us consider a situation in which we position the geometric focus of our lens, of the microscope objective, in a plane that we call j (Fig. 5.4).[3]

Browsing the sample, moving the focus of the lens, what we observe when we are in plane j, Oj, can be summarized by writing: $Oj\ Ij \cdot Hj + \sum_{k \neq j}(Ik \cdot Hk) + N$.

Here $_{Ij \cdot Hj}$ is the true information we are interested in by focusing the lens in plane j, where Ij is precisely the true distribution of fluorescent molecules in plane j distorted by the microscope according to its characteristic Hj. The summation of the contributions of all the planes k, with $k \neq j$, in which we can think of "slicing" the sample, $\sum_{k \neq j}(Ik \cdot Hk)$,

[3] A. Diaspro et al., *Three-dimensional representation of biostructures imaged with an optical microscope*, "Image Vision and Computing," 8(2):130–141, 1990.

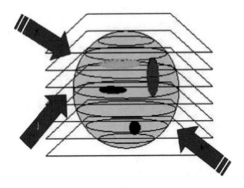

Fig. 5.4 Imagine the various details you want to observe confined in a sphere that is ideally "sliced" by parallel cutting planes. Let plane *j* be the one where, at a certain instant, the geometric focus of the lens resides, and let planes *k* be the ones above and below the one of interest that send information not required at that instant

makes the observation in a certain plane collect "fuzzy" information coming from adjacent planes (Fig. 5.5). *N* is the noise, the unexpected signal, considered additive and therefore treatable experimentally or theoretically, and that we neglect to simplify the treatment.

So, when you look at something under a microscope, the simplest microscope, at some point, while "blurring", you say "I am on fire". Aside from a possible connection to Bruce Springsteen's *I am on fire* (1984), the phrase should "sound" like this: "I have placed the focus of the objective in the area of best contrast, in the position where I 'see best'." If you consider what you observe in plane *j*, *Oj*, you may agree that when the contribution to observation that comes from the plane of focus "wins" over the contributions from the planes adjacent above and below, *k*, then you will be able to see a sharp image. If we assume that the point sources are homogeneously distributed in the cell you are observing, here is the reason why you

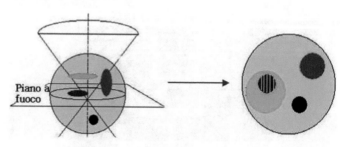

Fig. 5.5 The effect of the various information present in three-dimensional space is to present itself to our eyes or to the sensor of our microscope all at once

generally place the focus in the equatorial plane, in the center of your cell, where the number of point sources are greater than in other cutting planes. It matters little to us whether we are in the center or not. What we do is collect images at all planes and then use good mathematics to solve a basically simple problem. We need to trace the distribution of point sources for each I by relying on knowledge of O, which we observe, H, which we can measure or model, and N, which we can determine.

It will be a matter of solving, with a computational tool such as the electronic-digital calculator on which you are recording the images, a system of linear equations that will lead, for each plane in which you have positioned the focus of the lens, to determine the best estimate of the true distribution of the fluorescent molecules with which you have "marked" the biological molecules of interest. The problem can be solved in three coordinates, or plane by plane. At the end of the plane-by-plane reconstruction we have a three-dimensional map made as if for each plane we had been able to acquire the best image. The operations to be done are not error free and, in some cases, could introduce artifacts. The sequence of operations involves a chain of this type: acquisition of the images in space (x, y, z), application of the Fourier transform, calculation to determine the best estimate of the true image plane by plane, and application of what is called the inverse Fourier transform that allows us to bring back the solution found in terms of spatial frequencies in the more usual coordinates (x, y, z). This method is called computational optical sectioning and can be performed with any optical microscope simply by moving in a controlled way the focus of the objective plane by plane and recording each resulting image, from the sharpest to the most "blurred".[4] Each image contains useful information for the solution of our system of equations and the calculation allows us to use it.[5]

What problem does the processing and the sequence of images that we acquire by "blurring" solve? They give us the possibility of extracting three-dimensional information from a series of two-dimensional views by eliminating, plane by plane, "postcard by postcard", the nebulosity due to blurring (Fig. 5.6).[6]

Sometimes, the excess information collected, compared to that desired, does not allow us to make decisions about what we are observing. It does not allow us to understand the situation around us.

We may find ourselves in the situation of the nautilus, a tetrabranchiate cephalopod mollusk with a primitive visual system. The nautilus has a simple nervous system, with rather elementary calculation and reasoning capabilities. Should a predator approach, if the nautilus were equipped with a "normal" visual system, it would have to deal with

---

[4] K.R.Castleman, Digital Image Processing, Pearson/Prentice-Hakk, 1979.

[5] A. Diaspro et al., *cit.*, "Image Vision and Computing," 8(2), 1990.

[6] https://svi.nl.

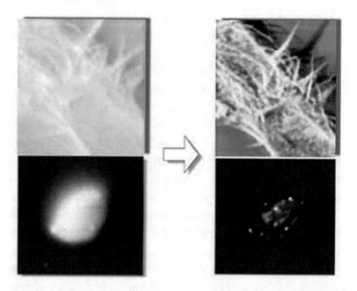

Fig. 5.6 "All it takes is a little bit of sugar..." would sing Ju-lie Andrews in the evergreen *Mary Poppins* (Robert Steven-son, 1964). A little calculation is enough to extract the information hidden in three-dimensional space, would say Hans van der Voort, a Dutch physicist

a set of images of the predator that make the scene rather complex. This would not allow it to "escape" in time and would become prey. For this reason, we like to think, nature has provided it with a pinhole—a small hole— which makes it see a simpler scene and makes it recognize more easily the approach of the predator. Simplicity and the need to handle an optimized set of data to make deci- sions, in this case it is realized through a physical selection of the information itself.

Let's try to see how we can realize a selection of the information we collect while we move the geometrical focus of the lens. We set ourselves the problem of elim- inating out-of-focus information, coming from the k

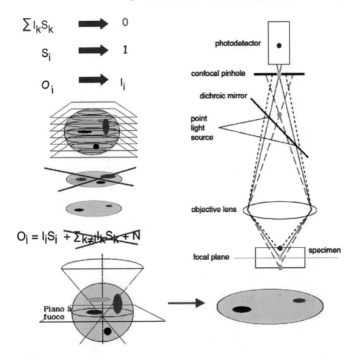

Fig. 5.7 The Columbus' egg, a bright idea: selecting the light rays coming from the *j plane of* focus by physically eliminating those coming from the *k* planes. An opaque screen with a small hole is all that is needed to perform a spatial selection of the information coming from our three-dimensional sample. The information previously "mixed" in the plane of focus is separated. By shifting the focus of the lens each cut plane provides an image as if it were the only plane existing "confocal" with the observation

planes, when we are interested in the signals coming from the *j* plane only, the one where we put the geometric focus of the lens from time to time. Figure 5.7 shows a synthesis of the situation in which, in front of the illumination, we obtain a signal, the fluorescence in our case, coming from the focus plane *j* and the adjacent planes *k*, that reaches the sensor that records the images, the substitute of our eye, in the observation of what happens beyond the lens.

This happens because in sending illumination at a certain focus point of the lens, with an appropriate wavelength to trigger a certain fluorescence, we cannot help but excite all the molecules encountered by the illumination beam, above and below the plane of focus. Our goal is to eliminate the signal coming from the points above and below the point in focus. To do this let's try simply tracing the light rays from the point emitters till they reach the sensor. Rays coming from points above or below the lens focus have their focus point before or beyond the sensor. So, the sensor would receive that information, the *k-plane* information, out of focus, together with the in-focus information coming from the *j-plane*. If before the sensor we put an opaque screen that blocks all light rays except those coming from the *j-plane*, by letting them pass through a small hole that, technically, is called a pinhole, then we can make the sensor receive only the information of the light point of the plane of focus. To realize the entire "postcard", the two-dimensional image, it is enough to move the illumination and the screen with the *pinhole* point by point. The images we collect, moving the focus of the lens from one plane to another, are those related to the only true information in that plane. Not considering the noise N for the reasons already said, we obtain a situation of the type $Oj\ Ij\cdot Hj$.

From each plane, due to the *pinhole* effect, we obtain the "true" image only affected by the limit imposed by diffraction. Each plane offers the observer the image "in focus", with the immediate and evident advantage of being able to observe better the details and to remove that "patina" of blur that hides them (Fig. 5.8).

We are pointillist painters using a finer-tipped brush, allowing us to bring back those point sources that make

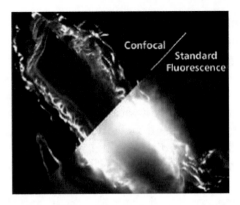

Fig. 5.8 Details previously hidden by the clutter stand out obvious floor by floor in the confocal version

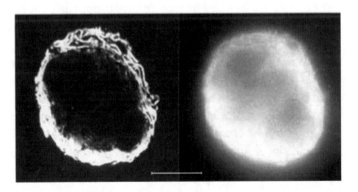

Fig. 5.9 The specimen is a plasmacytoma cell, malignant plasma cell tumor, labeled with anti-endoplasmin, which binds primarily to the cisternae of the endoplasmic reticulum. In the conventional image, it is not possible to determine whether the central nucleus is devoid of endoplasmin and the individual cisternae are unclear. Scale = 10 µm[7]

up our image more closely both in the plane of observation, and for each plane of our three-dimensional object.

It is a "historical" image (Fig. 5.9) that allows us to understand on one hand the effect of the confocal treatment and on the other to underline that the method should be used when a sea of fluorescence really prevents us from grasping details, even when performing a conventional optical sectioning.[7] The confocal microscope therefore allows us to make three-dimensional images in which we can navigate, perhaps with augmented reality glasses.

Due to the effect of confocality and of point by point interrogation, according to Giuliano Toraldo di Francia vision, we obtain a better spatial resolution both in the xy plane and along the z axis.

Giuliano Toraldo di Francia's idea is basically simple: the limit imposed by physics cannot be violated, but if in the process of image formation, of realizing our pointillist picture, we have additional information, then we can access data beyond the "resolving power" of the physical image-forming system, in our case the microscope.

Questioning matter "point by point" provides that "more" that direct observation does not offer. We know from which point in the real world, from time to time, the information comes. Or rather, we know what information, which would be confusing, to take away. We do not have this additional information when we observe the scene globally. In the computational case the additional information, not so powerful as in the confocal case, comes from the fact that the mathematical solution concerning the recovery of the best possible estimate of the real, being an intensity, can only be positive. In short, every little, or big, piece of additional information contributes to improve our pointillist picture, enriching it with details. The confocal microscope "isolates" the useful information

---

[7] W.B. Amos, J.G. White, *How the confocal laser scanning microscope entered biological research*, "Biology of the Cell," 95:335–342, 2003.

to which it adds the knowledge of the region from which they come.

With the confocal microscope, unlike computational optical sectioning where every photon collected is used for computation, the image formation process involves physically eliminating a portion of the signal generated by the excitation process.

However, to obtain a situation like $Oj\ Ij{\cdot}Hj$, we can use a method that exploits one of the revolutions, one of the most important paradigm shifts in the history of optical microscopy of living beings.[8] We will discuss this in more detail later, but now we are interested in understanding how this method can enhance access to information from the world in three dimensions.

This is the so-called two-photon microscopy. More precisely, the method concerns the excitation of fluorescent molecules, the ignition of our point sources, in an extremely localized manner in three-dimensional space using a nonlinear interaction.

The contributions $\sum \neq (Ik{\cdot}Hk)$, due to the two-photon fluorescence excitation mechanism, are set to zero, i.e., the fluorescence is excited only in the so-called focus volume (Fig. 5.10), where the photon density is maximal. Region that coincides with the PSF of the microscope that uses it.

Our "brush", as pointillist painters, immediately becomes thinner, as in the case of confocal microscopy.[9]

The excitation with two or more photons, foreseen towards the end of the Twenties, sees its origins in quantum mechanics and its experimental realization thanks to the advent of laser-type lighting sources (*Light*

---

[8] A. Diaspro (ed.), Confocal and Two-photon Microscopy. Foundations, Applications, and Advances, Wiley-Liss, 2001.

[9] A. Diaspro (ed.), ivi.

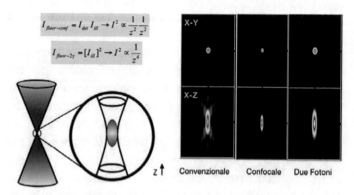

$$I_{fluor-conf} = I_{det} \, I_{ill} \rightarrow I^2 \propto \frac{1}{z^2} \frac{1}{z^2}$$

$$I_{fluor-2f} = [I_{ill}]^2 \rightarrow I^2 \propto \frac{1}{z^4}$$

X-Y

X-Z

Convenzionale   Confocale   Due Fotoni

Fig. 5.10 An alternative way to select a region along z is to use a nonlinear excitation mechanism, two-photon excitation. In this mode, only fluorescent molecules that are in a close neighborhood of the geometric focus of the lens are excited. The confocal mode works as if two lenses multiplied their signal-confining property, just as in the nonlinear case where the actors were two photons or more. The effect in both cases is to realize a pattern where the unwanted intensity along z decays with the fourth power of the distance from the focus point, i.e. abruptly

Amplification by Stimulated *Emission of Radiation*), that is, sources with "light amplification by means of stimulated emission of radiations", which had in Italy Orazio Svelto among its most important representatives and pioneers, particularly in the field of solid-state lasers and ultra-short optical pulses.[10]

To avoid collecting unwanted content in a certain region of the sample of our interest we can use a mode that I like to call "athletic". Typically, the microscopist seems to be "lazy" during the observation: he sends the illumination along a direction, which is generally referred to as the direction of propagation of light or z axis, and

---

[10] O. Svelto, *Principles of Lasers,* Springer, 2010 (1976).

always along that direction collects the light messages returned by the interaction of light with matter. Inevitably, as we have just seen, the contributions coming from the whole volume crossed by the illumination reach the eye or the sensor, creating that figure of confusion that we have learned to eliminate using calculus, a *pinhole* or two-photon excitation.

In the "athletic" mode, the microscopist positions himself, i.e., places his eye or sensor, at a right angle, at ninety degrees, to the direction of illumination. In doing so, what is observed is what happens in "a sheet of light" (Fig. 5.11). If you are dealing with a thick specimen, you have no contributions from out-of-focus planes, except for what happens within the thickness of the leaflet. A similar idea came to Richard Adolf Zsigmondy (1865–1929) who, in 1903 together with Heinrich Siedentopf, developed the ultramicroscope to better study colloids in solution.[11] It is an optical microscope with illumination not along the axis of view, but in which the observer receives only the light diffused by the observed corpuscles and not the direct rays coming from the light source. The effect is obtained by illuminating the object laterally or with a special lens called an ultracondenser. The ultramicroscope allowed Zsigmondy to conduct important studies on the colloidal state of matter, whose results made him a Nobel laureate in 1925. Among his most important studies were those on Brownian motions.

Returning to our light sheet, Fig. 5.12 shows its effect in the studio of tumor clusters or aggregates of cells well

---

[11] H. Siedentopf, R. Zsigmondy, *Über Sichtbarmachung und Grössenbestimmung ultramikroskopischer Teilchen, mit besonderer Anwendung auf Goldrubingläser*, "Annalen der Physik", 315:1–39, 1903.

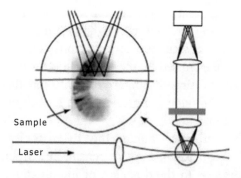

Fig. 5.11 Another way to select a region along *z* is to realize a "leaflet" of light that passes through the sample and is observed not along the direction of propagation of the leaflet but at ninety degrees. This allows the scene produced by the illumination to be observed without any contribution below or above the focus of the observation lens[13]

over 50 μm in size. When the sample is often a living organism, the advantage is even more evident.[12]

So, we've seen how, with our piece of glass and the light of the rainbow, it is possible to study living things in three dimensions without having to physically slice them up.

Exploring the living at low-level of perturbation we can pull together data from the molecular to the organ and tissue level to make a unique picture of "ourselves".

[12] J. Huisken et al., *Optical Sectioning Deep Inside Live Embryos by Selective Plane Illumination Microscopy*, "Science," 305:1007–1009, 2004.

[13] E. Stelzer, *Light-sheet fluorescence microscopy for quantitative biology*, "Nature Methods," 12:23–26, 2015.

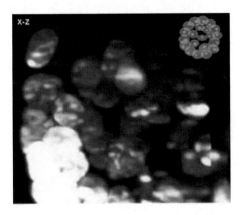

Fig. 5.12   The light-sheet scheme works particularly well in the case of thick samples such as embryos or this tumor spheroid, a model system for metastatic aggregates. *Credit* Zeno Lavagnino, DiasproLab, IIT

6

Modern Times: Space and the Time of Observations

*Everyone can take photographs: it's a pleasure, a way to express
oneself and fix memories, a way to explore the world
and be creative.*
Steve McCurry, photographer and photojournalist (1950

What happens when in the well-defined space of our body, in those boundaries that make us recognize organs and tissues, within our cell's DNA and proteins which move from one part to another, or when cells migrate from one area to another passing boundaries we thought insurmountable (Fig. 6.1)? The conquest or abandonment of spaces in time, the movement, gives us insight into not only what is happening, but can be important in predicting what will happen next. In the case of the possible development of a tumour, they can be important elements of pre-neoplastic diagnosis, or in the case of the brain premonitory signals of neurodegenerative diseases. They can offer those clues that allow us to apply a drug-logic solution in time or even better to see if it works. The

© Casa Editrice Ulrico Hoepli S.p.A. 2022
A. Diaspro, *Expedition into the Nanoworld*,
https://doi.org/10.1007/978-3-030-94472-8_6

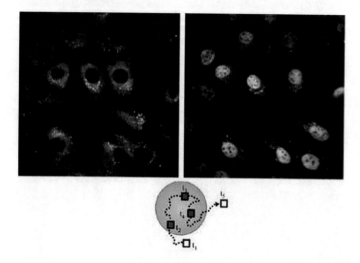

Fig. 6.1 Significant translocation of fluorescent molecules from one cellular compartment to another over time, from left to right

movement, the sequence in time of what happens in those spaces we consider usual, allows us to understand better, by observing. The light microscope is often used on samples, cells or tissues and organs, fixed. What does this mean? It means that in a very precise moment their functioning has been stopped, if we were in *the Disney movie Frozen,* we would say "frozen", which is what is done in the cryo-microscopy that has made the honor of the Nobel Prize for chemistry in 2017 to Jacques Dubochet, Joachim Frank and Richard Henderson. Cryo electron microscopy allows the three-dimensional structure of molecules to be reconstructed in atomic detail to the point where it is possible to visualize what "pocket" a drug goes into to be efficient or not. It involves purified proteins and fixed samples though a modality that "freezes" the living person to obtain a "still image" detailed in the extreme (Fig. 6.2).

The fixed sample, perhaps a sequence at different times but inevitably of, or by, different cells, can be finely

Fig. 6.2 3D model of the nucleosome based on cryo-microscopy data where the supercoiled DNA chains can be seen. The scale is that of a few nanometers

observed and can contribute to understanding the evolution of a "something" that proceeds in time. The painter in us could produce that pointillist painting we have referred to so many times and which will serve to make decisions, to draw conclusions. It could be a painting like Théodore Géricault's *Le derby d'Epsom*, 1821, oil on canvas, 91 × 122 cm, kept in Paris at the Louvre (Fig. 6.3). Looking at that painting, for years people understood that during a horse's gallop there is a moment, an instant, when all four legs are completely raised off the ground in the position of their maximum extension. Only a curious person like the American businessman Amasa Leland Stanford (1824–1893), director of the Central Pacific Railroad and founder of Stanford University in 1885, could come up with the idea of asking a great photographer, with a questionable life without "but", like Muybridge of verify what

Fig. 6.3 Fixing what happens temporally by confining it to space while time passes can lead to wrong conclusions

was happening. Muybridge's studies changed everything. They modified the way of recording observations and interpreting them. Muybridge with his *The Horse in Motion* (1878) is undoubtedly the pioneer of motion photography. Muybridge successfully photographed a running horse, setting up 24 cameras along the way by having the single camera activated by a "wire" struck by the horse's hooves and demonstrating that the hooves rise from the ground at the same time but not all in full extension (Fig. 6.4). Everything changed, Edgar Degas drew inspiration for his drawings, and Paul Valéry noted how Muybridge's photographs revealed the errors of sculptors and painters in representing the different gaits of the horse (*Degas Danse Dessin*, 1938).

For the observation of the living cell and all the movement of biological molecules, time, movement in space, is at least as important for understanding what is happening, and better still, what will happen!

If we think about the light in our rainbow, it's worth knowing that the time of illumination, the time in which we send photons into space, can be controlled at a very small scale, down between picoseconds and nanoseconds,

Fig. 6.4 A "rapid" temporal sequence of what is happening in space provides more accurate information by allowing us to grasp spatio-temporal details, i.e. movement

between thousandths of a billionth of a second and a billionth of a second. Modern lasers are not just "powerful" but ultrafast. We can administer light by precisely knowing the cadence of illumination. As with the distance of thunder, we only need to count to get precise information about the position of the light emitters by measuring their return time, a kind of echo. Also with light, we can trap living cells for a while, keeping them that way. The 2018 Nobel Prize in Physics was awarded to Arthur Ashkin, Gérard Mourou and Donna Strickland for their research that revolutionized the physics of lasers and light sources precisely by controlling the time and over time propagation of waves or photons in space paving the way for applications in many fields, including biology. We can use time to derive a lot of information in space, from the precise point from which light arrives—we will see this for super eyes—to the mobility or otherwise of the molecules we are studying. We can derive information at the level of molecular distances, on a scale of a few nanometers apart, using time or the different colour of the emitted photons. In short, by shuffling the cards we can increase the informational power of our piece of glass when it encounters light, light in its diversity. And as diversity gives value to life, light with its diversity increases even more the impact of the optical microscope in the study of life and the living.

Here we are in the world of augmented dimensions, we are in the fourth dimension! Have you ever run your fingers through, quickly, those little animated books that give you the sensation of movement and that children like so much? We are well beyond that!

We are particularly interested in understanding how proteins move and how molecular complexes are organized in space and time. How do cells communicate with each other, how does information propagate between neurons over time, how long does it take a protein to cross a cell or RNA to carry commands to make new proteins to ribosomes.

We can do a simple experiment, at least in its starting elaboration, to see how proteins move in the cell, to observe molecular trafficking. We take advantage of what we already know and can control. We can use the lens to send a certain illumination to a precise region in the three-dimensional space occupied by our cell. Typically, we trigger the phenomenon of fluorescence and receive photons in exchange for excitation. Fluorescent molecules, after a certain number of stresses, we know stop getting excited, i.e. they stop returning photons. You could stop reading but maybe, after a while, after a walk or a good coffee, you will resume reading. Fluorescent molecules do not. Their classic mode is to switch off forever.

The phenomenon we have seen is called *photobleaching*: the light that excited them makes them "mute" forever. Basically, this happens even among living people, in the best families or communities. Sometimes for someone who goes away, someone else comes along, or not.

Let's observe the fluorescence of the molecules we have labeled, let's suppose we want to know how fast or slow molecules pass through a certain area of the cellular universe. We point our illumination at that area causing the quenching of the fluorescent molecules by *photobleaching*. When our illumination is very intense, we increase the number of times we trigger the excitation-emission phenomenon by "getting tired" the molecules quickly. We call the instant of shutdown "time zero" and the associated fluorescence Fb, called *bleaching*. At time zero the intensity of fluorescence that, until a moment before had a certain value, $F0$, collapses to zero or more precisely to a value that we consider zero.

If we observe the quenching region, we can witness at least two phenomena that can be easily classified as distinct (Fig. 6.5). The first one is the one for which the intensity in that area remains zero. It means that the molecules we turned off were "immobile" molecules, that is, molecules that, for some reason, were anchored to specific structures in that region and their place could not be taken by any other molecule. The second one concerns the fact that, in time, fluorescence in the region where it was switched off starts to grow up to a certain value, $F\infty$. Typically, $F\infty$ does not exceed the initial fluorescence value, the one measured before switching off. If the initial value is completely reached this means that the molecules were all mobile. Otherwise, we will define the percentage of immobile and mobile molecules just by the intensity value reached, since we know the linear relation that links the fluorescence intensity to the number of molecules

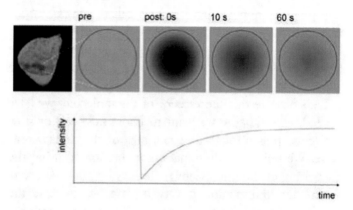

Fig. 6.5 Considering an area in a cell, observe the signal with its fluorescence intensity (pre). Increasing the excitation produces an irreversible loss of fluorescence in the area after some time, and the fluorescence intensity drops to zero (post). This process is named *photobleaching*. From the zero instant the signal is recorded, for example from 0 to 60 s. In time, we see a sudden decrease of the signal that will tend to grow again due to the movement of fluorescent molecules. The initial intensity can be recovered or not; this is FRAP (fluorescence recovery after photobleaching)

(Fig. 6.6). If we define the mobile fraction k as $k = (F\infty - Fb)/(F0 - Fb)$ then the immobile fraction will be equal to $(1 - k)$.

Moreover, the knowledge and measurement of the area affected by *photobleaching*, and the time course of fluorescence recovery, allows us to estimate the diffusion D of the molecules. Hence, in the hypothesis of free diffusion of the molecules, the Stokes-Einstein relation, $D = kT/6\pi\eta rH$ allows tracing of interesting information of the involved molecules, such as the viscosity η, or the hydrodynamic radius rH, where k is the Boltzmann constant and T the temperature.

Here, under certain assumptions, we are able to say something about the diffusivity of matter D, a theme dear

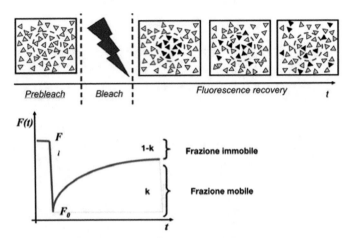

Fig. 6.6 If we call k the mobile fraction of molecules we can also calculate that immobility. We can understand if the molecules are free to move, if they are captured somewhere or if they are irreplaceable in that position. All this we can understand by analyzing what happens in space and time

to Einstein, who developed a theoretical approach and a model still used today.[1] We can tell if molecules are free to move, if they find barriers, or if they are subject to currents (Fig. 6.7). The key to the interpretation lies in the observation over time of what is happening and the relation to a possible physical model.[2] So, in the case of free diffusion, by estimating D we can trace the viscosity of the medium where the molecules move, their average size, the reaction times to a drug, the way a "flock" of molecules, of proteins, move into the cell to go do something. Warning.

---

[1] A. Einstein, *Über die von der molekularkinetischen Theorie der Wärme ge-forderte Bewegung von in ruhenden Flüssigkeiten suspendierten Teilchen*, "An-nalen der Physik", 17:549–560, 1905.

[2] D. Axelrod et al., *Mobility measurement by analysis of fluorescence photo-bleaching recovery kinetics*, "Biophysical Journal," 16:1055–1069, 1976.

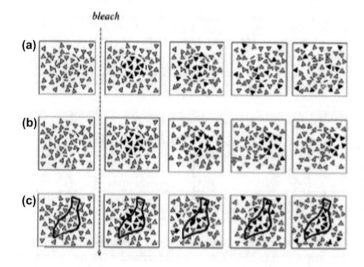

Fig. 6.7 Diffusion can be free, confined or find barriers. Information we can add for example to the relationship between the molecular weight of the molecule and its diffusion time

A cell is approaching! We are close to a surface! Here are the adhesion proteins "leaving" and going to settle, like diligent sentinels, where they are needed. So many things! We can already learn so many things from a simple experiment, from the simple perturbation we have triggered in that set of molecules we had made recognizable thanks to the fluorescent flag. We can make our observation more accurate by checking the times of perturbation, the volumes involved, the number of photons collected. Passing through our piece of glass with visible light we can learn so many things, so many more.

We can go further. We can scale up again, as Richard Feynman suggested, trying to reveal special properties of the molecules, typically fluorescent, that we want to observe. At the end of the day, it is a question of making considerations that link microscopy to spectroscopy.[3] Spectroscopy, the discipline that studies the way in which the intensity of radiation varies as a function of energy or "colour", was used by Sir Isaac Newton in 1665 to interpret the way in which light separated into colours when emerged from a prism, that old chandelier with grandma's crystal drops. The prism accounted for the energy spectrum of sunlight, those luminous bands that attract our attention as soon as a rainbow is born. So, we can "see" spectroscopically if two proteins, two molecules, that move in time in the confined space of our cells are more or less close with a precision that goes incredibly beyond the physical limit of diffraction, Abbe's law, Rayleigh and Nyquist criteria. It is something that the eyes just don't see unless we apply that "just observe!" to the spectrum of light that emerges from our samples. We can tell if two molecules are "close" in the order of distances between 1 and 10 nm, if you prefer between 10 and 100 Angstroms to emphasize that we are in a case where our observation, in colour, allows us to visualize approaches and departures of molecules in living cells at an atomic scale without violating the laws of diffraction.

The phenomenon that is exploited during the observation under the microscope is known as resonance *energy transfer* (RET) and is known in microscopy as FRET, assigning to the initial F the term *Fluorescence* or the name Förster. The chemist and physicist Theodor Förster (1910–1974) was the first to describe the phenomenon of energy transfer between fluorescent molecules.

---

[3] D. M. Jameson, *Introduction to Fluorescence*, CRC, 2012. P. P. Mondal and A. Diaspro, *Fundamentals of Fluorescence Microscopy: Exploring Life with Light*, Springer, 2013.

Between their fluorophores, making it possible to determine the relationship between proteins, lipids, RNA, DNA and other biological molecules on a molecular scale.[4] The description of this phenomenon is simple and colourful. Suppose we have at our disposal two fluorescent molecules called, referring to the generally used nomenclature, donor and acceptor. Two fluorescent molecules that we use to "label" the biological molecules that we are interested in studying in space and time. Let's state right away that the donor and acceptor can be used, for example, as flags to mark the beginning and end of a macromolecule. If this macromolecule breaks up, the donor and the acceptor move away. While if it compacts the two get closer, very close, within about ten nanometers. Taking the donor and acceptor individually, we know that if the donor is excited with blue light it emits green light while the acceptor, if excited with green light, emits red light. Those colours are the "optimal" absorption and emission λs for each individual molecule of the donor and acceptor. If the donor and acceptor get closer together, $d < 10$ nm, when the donor receives blue light, we no longer see green light but red light. What has happened? What has happened is that as a result of proximity, the donor has transferred the energy it would have used to emit green photons to the nearby molecule. This one, the acceptor, then used that energy to emit, in turn, the red photons that it would have emitted if it had been illuminated by green light. Fantastic! The proximity can be precisely evaluated using the efficiency of the process, the efficiency of FRET. The phenomenon somehow precisely exploits the electromagnetic wave-photon dualism. Electromagnetic

---

[4]T. Förster, *Zwischenmolekulare Energiewanderung und Fluoreszenz*, "Annalen der Physik", 437:55–75, 1948.

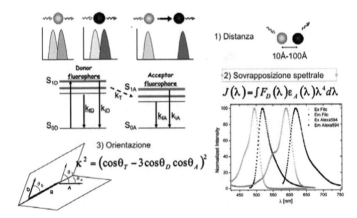

Fig. 6.8 When, for example, two fluorescent molecules approach each other, it is possible that when one is excited there may be a transfer of energy to the other which in turn produces fluorescence. Among the conditions for this to happen are distance, a certain overlap of excitation and emission spectra, and a certain orientation of oscillation modes. . Credit: Valentina Caorsi, Abbelight company, Paris, France

waves of appropriate energy bring the donor into the excited state but when approaching an acceptor is seen as an oscillating system that during oscillation captures the energy of the donor. Figure 6.8 schematizes the phenomenon and illustrates the characteristics of the actors responsible for theFRET.[5] The first of these is the distance, r, between the two fluorescent molecules. The distance determines the efficiency, E, of red versus green emission as the molecules in space and time approach each other, for example because the macromolecule whose beginning and

[5] R. M. Clegg, *FRET tells us about proximities, distances, orientations and dynamic properties*, "Journal of biotechnology," 82:177–179, 2002.

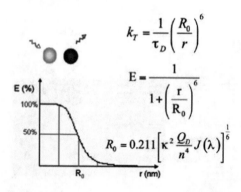

Fig. 6.9 The efficiency of this process, E, tells us how close the molecules are with a "strong" law that depends on the sixth power of the distance and a characteristic coupling parameter of the two molecules known as the Forster radius. *R0–R0* is the distance for which a specific pair of fluorescent molecules exhibits a 50% efficiency in transferring energy between them

end they mark curls up. The efficiency depends decisively on tiny distances, through the relation $E \sim (r/R0)^6$].

Förster studied and elaborated just the mechanism of this process and *R0 was* named Förster's radius. Now, how do we derive, from simple observations, the efficiency that allows us to trace the mutual distance between the two molecules? Simply we can say that the efficiency is measurable through the knowledge of how much emission occurs in the red spectral region compared to that which should have occurred in the green. To do this we can measure the green emission, generally that of the donor molecule before FRET coupling, which we call $_{FD}$. When the acceptor is present, which in turn will emit red light by energy transfer, we measure the remaining green fluorescence of the donor, which we call F_D. If F_D is zero in the presence of the acceptor, it means that all the energy acquired by the donor, instead of being transformed into green fluorescence, is transferred to the acceptor, and therefore the FRET process has an efficiency of 100%. The two molecules are very close and the rrelation $E - \Phi'_{D/FD}$ is 1, the efficiency is maximum. We note that if there were no energy transfer, the numerator and denominator would be equal, and the resulting efficiency would be worth zero. Figure 6.9 shows the trend of E as a function of distance. The trend of E in time allows, through the time sequence during which the acceptor donor molecules have approached, control of the interactions at a molecular level.

The excited donor emits fluorescence according to its typical emission spectrum. If the emission spectrum of the donor overlaps significantly with the absorption spectrum of the acceptor, the excited donor does not emit light but "transfers" the excitation to the acceptor in a, more or less, efficient way. The latter will emit a photon at its characteristic wavelength. The energy is transferred through so-called long-range dipole–dipole interactions.

What interests us is what is possible to be observed in the space and time of the observations. We have at our disposal a method that allows us to tell when, in a biological system, two molecules are approaching and when they are receding simply by observing the colour changes emitted by fluorescent molecules. You can imagine a great number of experiments and applications. Among all of them, you could check how efficiently a drug, marked by a fluorescent flag, can be detected or if it is able to reach diseased cells marked by another fluorescent flag waved by molecules characteristic of that disease. What makes this method so powerful is the fact that it can observe phenomena on a molecular scale, at the nanoscale, using a common laboratory bench microscope.

In addition to our piece of glass and the rainbow, we used stratagems and a good knowledge of the properties of our fluorescent flags to associate molecular level information regarding the diffusion and interaction of single

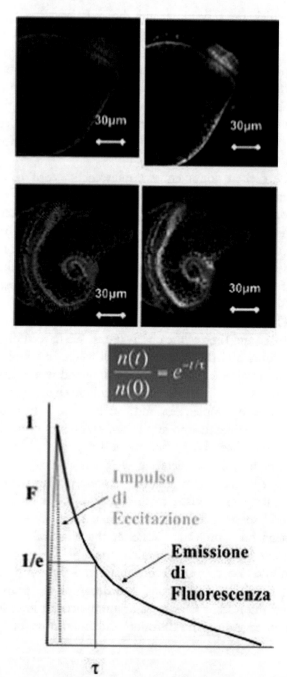

$$\frac{n(t)}{n(0)} = e^{-t/\tau}$$

Fig. 6.10 When a molecule or population of molecules is excited and is fluoresent, the emission of photons over time, $n(t)$, starting from an initial instant, $n(0)$, tends to wear out. It wears out more or less quickly-depending on the possibilities, in addition to fluorescence, that the molecule has to return to the resting state. By sending an excitation pulse it is possible to follow the phenomenon in time and to obtain what is called the average *lifetime of* the fluorescence. This time is strongly influenced by the boundary conditions. Thus, for the same emission recorded (boxes on the left), if a "lifetime" map is made, it is more detailed in relation to what happens in the vicinity of the fluorescent molecule. A healthy cell and a tumor cell emit a similar intensity of fluorescence from the nucleus but with different lifetimes to emphasize the different metabolic state

molecules in cells with microscope images without caring about the diffraction limit.

In addition to space, we have considered time and, as is often done, we have shuffled the cards, a bit like when we treat light as an electromagnetic wave or a photon. Time, the times, of molecules.

We know that fluorescent molecules have a time scale within which the emission of fluorescence is exhausted from the moment in which we illuminate them to excite them. The time scale we refer to is the nanosecond scale, and today we have sensors, electronic eyes, that allow us to measure arrival time of photons on that time scale.[6] What happens to time? We call this time the fluorescence lifetime. When a fluorescent molecule is in the excited state, from which sooner or later it will decay to return to the basic or quiet state, it can get rid of the excess energy that excites it by emitting photons or, in a non-radiative way, by thermally dissipating or by interacting with the surrounding environment, for example. The more the first

---

[6] M. Castello et al., *A robust and versatile platform for image scanning microscopy enabling super-resolution FLIM*, "Nature Methods," 16:175–178, 2019.

mode is prevalent, the longer will be the time during which it will emit photons. The molecule is not indifferent to what happens in its vicinity, so the lifetime of the fluorescence is influenced by the so-called boundary conditions. To be clear, if you received the same fluorescence in terms of intensity and colour of emission from nuclei of healthy and diseased cells and you went to measure the fluorescence lifetime, you would find that these are different because, for example, of the different first neighbors of the fluorescent molecule in the two cases. If you take the case of FRET, the perturbation induced by the approach of another molecule to the fluorescent molecule, be it donor or acceptor, changes the fluorescence lifetime. If you stain the DNA in a nucleus, you will find differences in fluorescence lifetime depending on the lesser or greater compaction of the DNA. There are several ways to measure and evaluate the arrival time of photons from the instant of illumination. We won't cover them here, but we will say that from these measurements we can derive lifetime maps that allow us to "see" differences that the intensity of emitted fluorescence alone could not pick up (Fig. 6.10). The method used to obtain these maps is called FLIM (*Fluorescence Lifetime* Imaging Microscopy): microscopy by images of the distribution of lifetimes. The times to which we refer are of the order of the nanosecond.

In the case of FRAP, we had considered fluorescence recovery over time to assess the diffusion times of biological molecules within the cell. These timescales are times slower than the nanosecond; typically, molecules diffuse on scales of the microsecond, millisecond or second.

In the time of diffusion of biological molecules, we can still learn a lot by observing molecules in motion with the method of FCS (Fluorescence *Correlation* Spectroscopy).[7] Imagine in a dark night you go to the window to keep it a little ajar and watch the headlights of the passing cars.

You can count them, you can count those lights that pass by the minute and in the event of an accident you understand it from those headlights that remain motionless or suddenly go out below at your windows. Here, we can do something similar by adding to the measurement of the diffusion of molecules the evaluation of their concentration and the fact that they can stay for a while, or forever, somewhere in the cell. Your ajar window will be a beam of light arranged, using the lens, somewhere in space. From a certain moment you start collecting the fluorescence signal, F(t), without moving. The fluorescent molecules passing by will make F(t) "fluctuate". From the distribution of the amplitude of the fluctuations and their duration we can derive molecular-scale information about what is happening in the living being as it occurs in the space and time of our observations.

---

[7] D. Magde, E. L. Elson, W. W. Webb, *Fluorescence correlation spectroscopy. An experimental realization*, "Biopolymers", 13:29–61, 1974.

7

Two-Photon Are Better Than One

The greatest show since the Big Bang is us, you and me.
Lorenzo Cherubini "Jovanotti", *Il più grande spettacolo dopo il Big Bang*,
in "Ora" (2011)

It's time to take a big leap forward with our piece of glass and the rainbow. To push us to the red border of the rainbow to gather information from the living without asking its cells to be placed between two slides. There is still much to be done, but the main road is taken. It is two-photon and not two photons because it is a single quantum event, as we will see later. Now, we can answer questions concerning cells when they aggregate, when they organize themselves in our organs and tissues, which we previously answered by placing cell samples on slides. The glass I am most familiar with is that of the beer mug, actually mostly that of the glass with which I keep my commitment to drink enough water during the day.

© Casa Editrice Ulrico Hoepli S.p.A. 2022
A. Diaspro, *Expedition into the Nanoworld*,
https://doi.org/10.1007/978-3-030-94472-8_7

From the cells on the slide, I learn many things but that is not the condition of the living being. Today, I can follow the molecules of a transdermal drug taken through the skin, just as I can follow the molecules of a cosmetic using red light instead of blue light to look at the DNA and the proteins of the cells by penetrating the light to a little more than a millimetre into the living being that we are. I can observe what's going on with the prospect of eliminating biopsies, of eliminating the need to take cells outside the human body to study them.

To do this I participate in a phenomenon that is wonderful, that leaves one breathless. It is the excitation of two or more photons. A beam of red light that, in spite of the rule that rightly follows the fluorescence to emit light signals in colours less energetic than those used for excitation, produces a beautiful green light. I still remember when, putting together the first Italian two-photon microscope in the laboratories of the Physics Department of the University of Genoa back in 1999, one Saturday morning I took Teresa to see that show, obtained the night before the weekly shopping trip. I remember the sleepless nights thinking about how to align the light beams and the meticulous use of that Trio Kenwood oscilloscope that my grandfather Mario had given me in 1978 when I was facing university studies with the romantic idea that at the University I would be taught to use it in the right way. Green fluorescence from red light, a bet in which the National Institute of the Physics of Matter, the legendary INFM, had believed when it informed me that the wise men of the Institute had decided to finance that research, the first and only one in Italy. I was to discover years later that Aldo Cingolani, Roberto's father, had been one of the pioneers of non-linear multi-photon interactions. The circumstance, coincidentally, was the beginning of the adventure at the Italian Institute of Technology to realize, this time, the first Italian microscope with unlimited optical super-resolution. It was 2009, and it was Roberto Cingolani who believed in the project, and we will discuss it in the next chapter.

We will therefore cover two-photon optical microscopy and its applications to the study of biological systems.[1]

The story of the two red photons instead of the blue one begins many years earlier when the doctoral student Maria Goeppert (1906–1972), later Goeppert-Mayer and Nobel Prize winner for physics in 1963, wrote her thesis, between 1927 and 1929, and entitled it "About elementary acts with two quantum jumps".[2] A courageous novelty about the theory of possible two-photon absorption by atoms. This theory is confirmed, not only because it is the result of a beautiful and elegant formulation as predicted by Paul Dirac,[3] but also because it made possible the discovery of a mechanism to produce light that changed "everything" in optics.

Let's go in order, what needs to be demonstrated first? The scheme envisaged by Maria Goeppert-Mayer (Fig. 7.1) concerns the fact that an atom or a molecule can be brought into an excited state through successive transitions. Imagine that to bring a molecule, for example fluorescent to get right to the point, into an excited state requires, in arbitrary units for convenience, energy 10. This means that that molecule, if illuminated by blue light, has a good chance to absorb it. The quantity we refer to in this case is called the cross section, or what area the molecule makes available to blue light to increase the

---

[1] A. Diaspro et al., *Two-photon fluorescence excitation and related techniques in biological microscopy*, "Quarterly Reviews of Biophysics," 38(2):97,166, 2005.

[2] M. Goeppert-Mayer, *Über Elementarakte mit zwei Quantensprüngen*, "An-nalen der Physik", 9(3):273–295, 1931.

[3] P.A. Dirac, Beauty as a method, Raffaello Cortina, 2018.

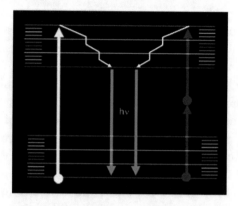

Fig. 7.1 It seems easy to say that to get somewhere you need energy 10, blue light, and you can get there in two jumps from energy 5, one suspended in the air like Cristiano Ronaldo and the other definitive, red light. It will take many years to prove it but once up there the photons emitted by the fluorescent molecule will be the same

probability of intercepting it. That specific molecule will make a much smaller area available for the other colours. Imagine on a Christmas shopping day that you are in the busy main street of your city or town. You see in the distance a person that you would like to meet and for this reason, not only because you are Italian, you reach out a little. This, increasing the area of your visibility, increases the probability that that person will see you, it does not guarantee that he will embrace you or that he will eventually notice you among all those people, but that was all you could do. If instead you don't want to meet someone you have glimpsed in the distance, maybe you make yourself small or you slip into some shop. The probability of the encounter follows the previous reasoning, it's not zero but it's very low.

So, our molecule is well prepared to absorb blue light, then it will act as a fluorescent molecule and emit in the green.

The two-photon absorption theory, more correct to call it double-photon as a quantum unit/entity—note that we use *two photon* and not *two photons for the* same reason—predicts that Sending, for example, two red photons, each carrying an energy of five, will produce the same effect as sending a blue one carrying an energy of ten. On the other hand, the sum of the energies makes the accounts correct.

Each of the red photons is not able to bring the molecule in the excited state. However, if the two are able to simultaneously intercept the same molecule there is a good probability of reaching the desired level of excitation by passing through what is called a virtual state, the "meeting point" of the red photons, those at lower energy in Fig. 7.1.

Try to imagine a classical situation that can be faced with simple school physics: a room, a ball, and a table. If the desk is high h and the ball has a mass equal to m, providing an energy equal to mgh, where *g is the* acceleration of gravity, you should succeed with a good probability to bring the ball from the ground onto the table. *The ball is on the table!* you would exclaim. If, however, you were to hit the ball with energy equal to $mgh/2$, an observer leaning tiredly against the doorframe of the room would see the ball at half height. "Tapping!" now, for in the room someone is arranging refreshments, putting delicacies on the table, giving a timetable to start the event. Anyone can participate as long as they approach the table giving "taps" of energy equal to $mgh/2$. You know as well as I do that the voracity of your friends is such that if you were to arrive late, you would find nothing or "secondary delicacies". At the indicated time, the observer leaning against the doorframe sees a lot of people around the table giving "taps" equal to $mgh/2$ and every now and then he notices that a ball at half height receives enough energy to be brought onto the table. This is how two-photon excitation of fluorescent molecules works. Lots of photons in the same place in a well-defined window of time. In fact, if you were on a sailing boat on a sunny day carrying a can of soda or fluorescent molecules you might see fluorescence from single-photon interactions, i.e. energy 10, approximately every second. But if you wanted to experience the phenomenon from two-photon absorption you would have to "wait" for about ten million years.[4] The reason for the long wait is that on one hand the cross section for a two photon event is very small compared to the single photon even. On the other hand the Maria Goeppert-Mayer theory predicts that the first photon, the first energy instalment, can be given back to your bank whenever you want. However, the second instalment, your bank or the molecule, has to claim it within the time of the absorption phenomenon, that is within 10^{-17} s! What the heck! Therefore, the event is very unlikely. How to realize it in a "human" time, for example in the time of our observation under the microscope? If you had ten million years at your disposal, you might want to do something else, too.

The phenomenon concerns the photons sent on the molecule to be excited. This being so, the best way to promote the phenomenon is to illuminate with as many photons as possible the region where you want to realize the two-photon excitation.

This is so true that it will take about thirty years for the experimental demonstration of Maria Goeppert-Mayer's theory. It was the realization of the ruby laser, the amplifier of light, in 1960 by Theodore H. Maiman of Hughes

---

[4] W. Denk, K. Svoboda, Photon upmanship: why multiphoton imaging is more than a gimmick, "Neuron", 18(3):351–357, 1997.

Laboratories[5] that paved the way for the demonstration of that non-linear phenomenon so rare in nature. The first experimental verification was in 1961 when Kaiser and Garrett, of Bell Laboratories in Murray Hill, New Jersey, recorded fluorescence from double-photon excitation in a crystal doped with europium.[6] In 1978 Sheppard and Kompfner obtained luminescence signals from an optical microscope and saw the possibility of applying the method to localize proteins.[7]

In order to make the phenomenon observable on a reasonable time scale, in both cases an attempt was made to confine as many photons as possible to the sample to be observed. The laser and optical lenses with high numerical aperture, those that make the brush very fine, were important for the so-called proof of *principle*, a sort of litmus test.

Obtaining images from living beings means finding ways not to destroy them during observation. The intensities involved are of the order of 50 gigawatts per cm^2, and if it is true that biological molecules absorb little radiation in the red region, it is equally true that absorption is not completely null. A small absorption disposition invested by an enormous number of photons produces effects. Before the experiment of Sheppard and Kompfner, Michael Berns had warned about the possible effects of two-photon excitation obtained with beams of illumination when they produce an intense radiation on biological

---

[5]T.H. Maiman, Stimulated optical radiation in ruby masers, "Nature," 187:493, 494, 1960.

[6]W. Kaiser, C.G.B. Garrett, Two-Photon Excitation in CaF2:Eu^{2+}, "Physical Review Letters," 7:229–231, 1961.

[7]C.J.R. Sheppard, R. Kompfner, Resonant scanning optical microscope, "Applied Optics," 17:2879–2885, 1978.

samples.[8] In 1990 the turning point: Winfried Denk, in the Cornell laboratories directed by Watt W. Webb, obtained images from biological samples, exploiting the emission of fluorescence obtained with two-photon excitation and making Colin J. R. Sheppard's observation and Maria Goeppert-Mayer's prediction more concrete.[9]

What has changed since previous attempts? The first pulsed lasers, pulsed on a decidedly short time scale, femtoseconds, fs, 10–15 s, burst into research laboratories. This changes the cards at stake because it makes it possible to administer, for a very short time, an enormous quantity of photons. Whoever receives them, if we wanted to use the power of the laser beam as a reference, "feels" a big bang of light for such a short time, typically 100 fs, that even triggering the excitation of two photons can not have time to change the structures. From Denk's article onwards, the community of microscopists develops variants and new microscopes highlighting the advantages of using red light compared to blue light to probe biological matter and more.

Since we know that fluorescence is an instantaneous phenomenon, it is necessary to continuously repeat the excitation in order to form an image. In the case of two-photon microscopy, the train of pulses to be repeated is modulated at a frequency of about 100 MHz. The choice of frequency is rational: it means repeating the pulses of excitation every 10 nanoseconds and since fluorescence is exhausted in about 1 nanosecond, there is no risk of "wasting" photons trying to excite already excited molecules.

---

[8] M.W. Berns, A possible two-photon effect in vitro using a focused laser beam, "Biophysical Journal," 16:973–977, 1976.

[9] W. Denk, J.H. Strickler, W.W. Webb, Two-photon laser scanning fluorescence microscopy, Science, 248:73–76, 1990.

From a point of view of the possible perturbation of the sample, it can be shown that the average power P_{av} seen by the system is reasonable, being valid the relation $P_{av} = D \cdot Ppk$, where Ppk is the peak power necessary to trigger the non-linear phenomenon while D is the *duty* cycle, the useful working cycle of the laser. In our case it can be calculated using the relation $D = \tau_p \cdot F_p$, where τ_p is the pulse duration and F_p the repetition frequency. This means that the very fast duration of the pulse, 10^{-13} s, is a kind of shelter from a possible photo-perturbation. As a final piece, red light, between circa 600 nm up to 1010 nm, is produced by modern lasers based on a titanium-sapphire crystal, beautiful to see and fantastic for its performance.[10] Under these conditions, the probability of bringing a fluorescent molecule into the excited state depends, at first approximation, on the following relationship using the number of photons absorbed per fluorescent molecule per pulse:

$$n_a \mathrm{H}\left(\sigma_2 P^2 / \tau_p F_p^2\right) \cdot \left(NA^2 / 2\hbar c \lambda\right)^2.$$

Leaving aside the temporal considerations, and recognizing the Planck constant, the speed of light and the wavelength, let us concentrate on the three key parameters: σ_2, P and NA. NA, the numerical aperture of the lens, matters a great deal, with its fourth power. This is reasonable because it represents the ability to concentrate the illumination, hence the photons. It is important that it is greater than 1, suggesting the use of an immersion medium, be it water or oil. The power P of the illumination tells how many photons are emitted per second. So P represents the probability with which the event can occur. Since the

---

[10] O. Svelto, *op. cit.*

"collision" of each of the two photons with the molecule is an independent event, the probability of simultaneous absorption of two photons, in that one quantum event described by Maria Goeppert-Mayer, goes with the product of the two probabilities that is with P, average, just to the second power.

Technically we are in the domain of high photon densities that we obtain with a high numerical aperture lens and a "powerful" illumination source.

We are ready to bring in the main recipient of this epochal revolution, the fluorescent molecule. We do so through its two-photon cross section, σ_2. The only way to know its value is to measure it, while it is impossible to predict it even with the most sophisticated calculation tools.[11] The most interesting property of the two-photon cross-section is that while in the case of single-photon excitation the single-photon cross-section selects a rather narrow "bell" distribution of excitatory colours, when switching to two-photon the possibilities are wider (Fig. 7.2). So much so that a two-photon wavelength can be used to excite multiple fluorescence including intrinsic fluorescence or autofluorescence, which are classically activated in the single-photon regions of violet and blue. The selection of the different emissions will be done in registration of the arrival of photons, energy by energy, colour by colour. The diagram in Fig. 7.3 shows that when the molecule arrives in the excited state, it does not matter, for the continuation, if it arrived there by excitation by one or multi-photon.

[11] C. Xu, W.W. Webb, *Measurement of two-photon excitation cross-sections of molecular fluorophores with data from 690 to 1050 nm*, "Journal of the Optical Society of America B", 13:481–491, 1996.

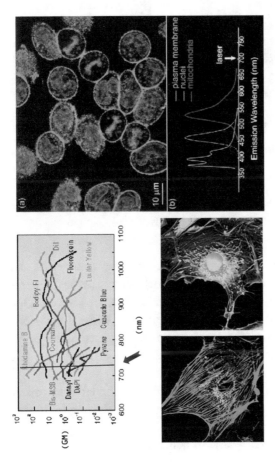

Fig. 7.2 The propensity to be excited by two photons is measured in GM, in honor of Maria Goeppert Mayer, colour by colour. What you find yourself mentally hoping for is that, unlike the single-photon case, we don't have a bell within which to excite but something "unraveled" that allows for a winning pattern. With the light of a single colour we can simultaneously excite several fluorescent molecules of which we will be sure that they are all in that small volume of two-photon excitation that turns out to be about 0.1–0.01 femtoliters

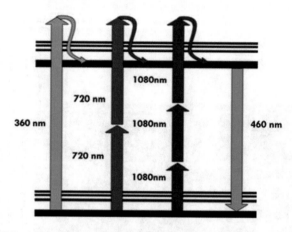

Fig. 7.3 Instead of using energetic radiation from the violet-blue region of the spectrum, you can use colours with less impact towards red and make two- or three-photon schemes to bring molecules into the excited state. Whatever the mechanism by which the molecule is brought into the excited state, its mode of fluorescent emission does not change

The main advantage lies in arriving there with an excitation in the red region, which allows better penetration into large cellular clusters, organs, or tissues. Due to the spatial selection of the event in three dimensions (Fig. 7.4), once the fluorescence is triggered there is no doubt that it can only come from the region where the geometric focus of the lens is.

It's time to turn off all the lights in the lab and collect each photon per emission energy, by assigning it to that point in your pointillist painting.

We are therefore able to make three-dimensional images by optically dissecting our sample without the need for a *pinhole*. In this way we significantly increase what is called the signal to noise ratio.

We can obtain three-dimensional images, as in the case of computational optical sectioning or confocal

Fig. 7.4 We thus illustrate the spatial selection of two-photon excitation. At single-photon, 1PE, all the molecules encountered in the excitation path are likely to emit fluorescence while in the two-photon scheme, 2PE, only the molecules that are in the region of maximum photon–ca density emit fluorescence, for the others the waiting time is about 10 million years, although the event is unlikely at low photon densities. The photograph shows the fluorescence, top, coming from conventional illumination and the point fluorescence coming from two-photon excitation. The advantage, in 3D space, is obvious

microscopy, and penetrate "thick" systems such as the first layers of the skin. Imagine that you can more efficiently track cosmetics or transdermal drugs in the skin. The high ability to confine the interaction to precise regions of the sample allows us to design new experiments where the illumination beam can be actively used and to realize three-dimensional interrogation schemes with globally reduced *photobleaching*.

Suppose you are studying a cellular system of which you would like to know how the information contained in the nucleus is propagated to subsequent generations. In this case, if you have cells in which the DNA has been labelled with photoactivatable fluorescent molecules, and you have found the two-photon photoactivation threshold,[12] you can photoactivate the DNA under two-photon conditions in the nucleus of a "mother" cell and when you see fluorescence in subsequent generations, it can only have propagated from the areas that you had photoactivated in the "mother" cell. On the other hand, if you take a fluorescent sphere, you realize that with two photons you can "write" in a localized way by exploiting the fact that even *photobleaching*, which can be considered an extreme photo commutation phenomenon, can be localized. In both cases it would not have been possible to realize the same event selection using a single photon, simply because photoactivation and *photobleaching* could have been triggered at any point of the double cone that makes the geometry of the lens, where the energy would have been directly appropriate.

---

[12] M. Schneider, A. Diaspro et al., *Two-photon activation and excitation properties of PA-GFP in the 720–920-nm region*, "Biophysical Journal," 89:1346-1352, 2005.

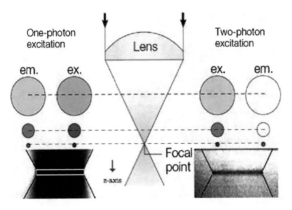

Fig. 7.5 When we deal with fluorescent molecules, we know that after a certain number of excitation-emission cycles they stop emitting forever, they wear out. At single photon we consume them throughout the excitation volume regardless of whether or not we use all of the emitted fluorescence to form an image. In fact, if we excite a line along the *z-axis* and we go to see what happens when *photo bleaching* occurs in that line, we realize that even above and below the molecules are consumed. At two photons instead we consume only what we observe, leaving unchanged the fluorescent molecules in the adjacent planes that we can then interrogate by moving the focus of our lens along *z*

Another aspect, no less important than the previous ones, concerns the consideration of the fact that during the acquisition of three-dimensional data, when you move the excitation from one plane to another, unlike the single-photon case that photo-perturbs molecules above and below the plane of interest, under the two-photon regime the *photobleaching* results are spatially confined (Fig. 7.5).

Thanks to the high photon densities involved while triggering fluorescence, it is possible to produce light signals, spatially confined, exploiting the phenomenon of

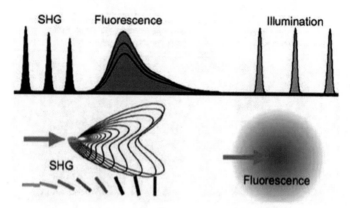

Fig. 7.6 The intense illumination used for two-photon excitation makes it possible to trigger an extremely interesting phenomenon, "for free". Some structures, collagen for example, are able to transform the intense red light into blue light, i.e. to double the oscillation frequency of red. This phenomenon of second harmonic can be added or not to the fluorescence offering complementary information, typically along the direction of propagation

second harmonic generation, *SHG* (*Second Harmonic Generation*).[13]

If the sample under observation is in a particular structural situation, for example in what is called breaking of the centres of symmetry, the high photonic densities are able to generate a signal in the forward direction, that of the illuminating beam, called second harmonic because its wavelength is half of the illuminating one, i.e. blue is generated by red (Fig. 7.6). Although the second harmonic signal is coherent and its propagation occurs in the "forward" direction, it is possible to collect the SHG signal

---

[13] P. Campagnola et al., *High-resolution nonlinear optical imaging of live cells by second harmonic generation*, "Biophysical Journal," 77:3341–3351, 1999.

also in the backward direction,[14] the one that is generally used to record fluorescence. The information that can be gained from simultaneous three-dimensional recording of forward, backward and autofluorescence information[15] has opened up fascinating and effective prospects in what is called *label free* microscopy: optical microscopy without added markers.

---

[14] A. Diaspro et al., *Time-variant analysis of organelle and vesicle movement during phagocytosis in Paramecium primaurelia*, "Proceedings of SPIE", 4622, 24–31, 2002.

[15] P. Bianchini, A. Diaspro, *Three-dimensional (3D) backward and forward second harmonic generation (SHG) microscopy of biological tissues*, "Journal of Biophotonics", 1(6):443–450, 2008.

8

Super Eyes to Watch Light Signals

No limits Jonathan!
Richard Bach, *Jonathan Livingston Seagull*
I try to believe there always is a brighter spark in the dark so stark
Premiata Forneria Marconi, "*If I had wings*" in *I dreamed of Electric Sheep*,
lyrics by, M. Marrow; music by F. DiCioccio, L. Zabbini, P. Djivas, 2021

With the best microscope available in research labs, until the advent of the super-resolution that makes our eyes super eyes, nuclear pores, entry and exit ports for DNA and other molecules between the cell nucleus and the cell could be seen as in Fig. 8.1. After all, knowing how they are marked and knowing that the fluorescence comes from nuclear pores could tell you something, but not enough to count them, or to see their individual constituents. With super eyes the situation changes, and we see the individual nuclear pores, the proteins that make them up and possibly the traffic that passes through them (Fig. 8.2).

© Casa Editrice Ulrico Hoepli S.p.A. 2022
A. Diaspro, *Expedition into the Nanoworld*,
https://doi.org/10.1007/978-3-030-94472-8_8

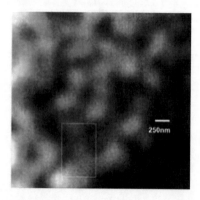

Fig. 8.1 Nuclear membrane pores observed with a state-of-the-art fluorescence microscope

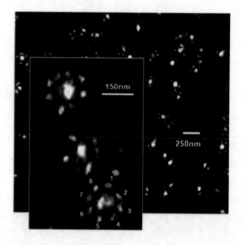

Fig. 8.2 Nuclear membrane pores observed with a super-resolution fluorescence microscope

Improving spatial resolution is almost an obsession for optical microscopists breeding tension that sees them, sees us, competing with the electron microscope. Forming better spatial resolution images is not just about detail.

If you look at Jean-Etienne Liotard's painting *La belle chocolaterie* (1745) and the details, you will hardly change your mind about liking it or not (Fig. 8.3). What about *Gala Contemplating the Sea Mediterranean by* Salvador Dalí *that at twenty meters becomes a portrait of Abraham Lincoln* (1976) or a Marilyn Monroe photographic shot, seen at low resolution, that becomes an Albert Einstein picture (Fig. 8.4)? This is the real world offered to our observation in different ways, a representation in images by observing which we will make decisions. It is not a matter of details whether a doctor can access molecular information or not to make a diagnostic or surgical decision.

Today we can really think of our optical microscope as a radio on which, instead of selecting the radio station, we can choose the desired spatial resolution needed to answer the current question. I would go so far as to say that the tuning is limitless, we can already locate molecules with an

Fig. 8.3 Detail is important but not everything. It imports the information content of every single point in the image

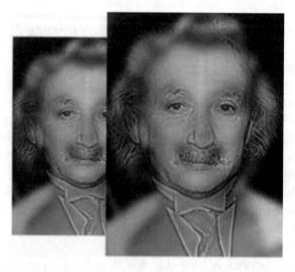

Fig. 8.4 The distance between Marilyn Monroe and Albert Einstein lies in the way our brains process the presence or absence of detail coupled with abundant pre-existing knowledge of the faces of the two decidedly popular characters

accuracy of 1 Angstrom or make images on a scale of a few nanometers.

Our eyes become super-eyes and we call this ability super-resolution, meaning that we can pick up anything beyond the usual, beyond the limits imposed by the physics of our piece of glass and of the rainbow light to which we refer in lighting and signal detection.

The term super-resolution was introduced into the international scientific community by a brilliant Florentine physicist, Giuliano Toraldo di Francia (1916–2011). Toraldo di Francia, when asked if it was possible to exceed the resolution limit synthesized by Abbe's law, said yes. Violating perhaps the laws of physics? No way. It is enough to add other information, known for whatever reason, to those captured with our lens to form an image and we can go further.[1] It is a matter of going around the diffraction limit rather than trying to violate or break it down.[2] Just add "a priori" information about the system you are going to study, and you are done! If you see a doodle on a sheet of paper and nobody tells you anything, that doodle remains a doodle, mostly it remains incomprehensible to what it might hide. But, if someone tells you that he has drawn two dots close together on a sheet of paper and proposes it to you again, here is the form of that squiggly thing might make you wonder where the two dots are. Poorly drawn, of course, but present on that piece of paper.

What additional knowledge can we bring into play to decipher what comes to our eyes confusedly, for example, as a result of the laws of physics that through diffraction proposes to us two very close light emitters as a sort of luminous doodle? Which and how many arrows do we have in our quiver?

If we use computation and face the problem of reconstructing a "true image" starting from what we can observe, the only constraint of the positive solution combined with the knowledge of a reliable model of the image forming system of the optical microscope, in our case, allows us to improve the spatial resolution of the image we try to recreate as we have seen in reconstructions in computational optical sectioning microscopy.

I can mix known and unknown details of the sample I am studying and of which I would like to obtain a distribution of particular molecules beyond the limit imposed by diffraction. Mats Gustafsson (1960–2011)

---

[1] G. Toraldo di Francia, *Resolving Power and Information,* "Journal of the Optical Society of America," 45(7):497–501, 1955.

[2] A. Diaspro, *Circumventing the diffraction limit,* "Il Nuovo Saggiatore," 30(5):45–51, 2014.

has successfully proposed something similar, interpreting very well the suggestion of Giuliano Toraldo di Francia with his structured light microscope.[3] The structuring of the illumination, obtained by inserting along the optical path reticles with known characteristics, in different positions, allows us to "load" them in some way with the information that cannot be deciphered with the simple microscope, thus preparing the set of data that allows us to reconstruct images in super-resolution (Fig. 8.5).

Spatial resolution can also be improved by distributing over several image points, organized with a known geometry, the information from each illuminated point reassigning, in some way, the apparently precisely diffuse collected signal.[4] Nowadays, this is called ISM, *Image Scanning Microscopy*, and from a pioneering experimental realization[5] it is at the heart of the microscopes of the future, offering both super-resolution and spectroscopic performances.[6]

On the way to super-resolution, an interesting method of preparing the sample to be observed has recently been used. The method is called "Expansion Microscopy".[7] A sort of "soufflé" of the sample to be studied is prepared. Following the specific marking in fluorescence, a procedure of fixation of the situation is carried out during

---

[3] M.G.L. Gustafsson, *Surpassing the lateral resolution limit by a factor of two using structured illumination microscopy*, "Journal of Microscopy," 198(2):8287, 2000.

[4] C.J.R. Sheppard, *Super-resolution in confocal imaging*, "Optik," 80:53–54, 1988.

[5] C.B. Müller, J. Enderlein, *Image Scanning Microscopy*, "Physical review letters," 104:198,101, 2010.

[6] M. Castello et al., *A robust and versatile platform for image scanning micro-scopy enabling super-resolution FLIM*, "Nature Methods," 16(2):175–178, 2019.

[7] F. Chen, P.W. Tillberg, E.S. Boyden, *Expansion microscopy*, "Science," 347(6221):543–548, 2015.

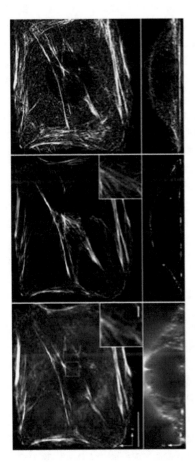

Fig. 8.5 The use of structured light produces additional information to the image formation scheme. Starting from the conventional image, super-resolved images can be computationally reconstructed

which a transparent polymer, a "gel", is made to bind in a precise and effective way, using excellent biochemistry, to the fluorescent molecules. Everything is removed from the original sample except the bonds between the fluorescent molecules and the polymer, leaving intact a sort of "crystal house" with the details we are interested in the right places. Then the sample, or what remains of it, is immersed in water and allowed to grow in volume. The effect of this preparation lies in the fact that the fluorescent molecules that before were too close to be distinguished by the microscope have been moved far enough away to be distinguished. It is thus possible to form an image that contains details that previously could not be grasped at distances smaller than the diffraction limit. But, after the "soufflé", they have become spatially distinguishable. Two of the most interesting aspects concern the fact that in this way the super-resolution can be obtained with any laboratory microscope and that the "recipe", which was not obvious, also works at the nanoscale, for example on nuclear pores (Fig. 8.6).[8]

In this panorama between stratagems and calculation strategies stand out the two methods that have led E. Betzig, S. W. Hell and W. E. Moerner to the Nobel Prize in 2014. With them the diffraction limit has crumbled!

The secret that unites the solution found by the three Nobel laureates is that to "go around", you could say "circumnavigate", the diffraction limit you've "just" got to find a way to preclude the simultaneous emission of two point sources, spectrally identical or that emit the same colour, when they are "too" close because they are at distances smaller than those allowed by the diffraction limit.

---

[8] L. Pesce et al., *Measuring expansion from macro- to nanoscale using NPC as intrinsic reporter*, "Journal of Biophotonics," 12(8):e201900018, 2019.

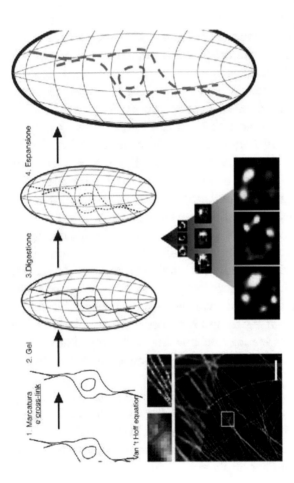

Fig. 8.6 The expansion process applicable on all scales is illustrated here. Even at the nanoscale of nuclear pores or cytoskeleton filaments

This solution is applicable to all those contrast mechanisms for which we are able to control two states such as light absorption and transparency, spin up or spin down, polarization in one state or in its opposite state. It is a general method that gives account to the idea of Giuliano Toraldo di Francia of being able to add information to the data that we can collect while limited by diffraction or any other physical law.

Betzig and Moerner find the solution to achieve super-resolution in fluorescence by requiring that image construction goes through consecutive viewing of groups of individual molecules that are positioned at distances greater than the Abbe distance. This makes it possible to distinguish them as individual entities and to locate them precisely by accumulating photons with the certainty that they always come from the same position.

The number of photons emitted, the quenching, and the direction of the oscillation of the dipole moment associated with the individual molecule are those elements that make us learn, molecule by molecule, to recognize individual molecules and map them in an image. This ability is the basis of Betzig and Moerner's super-resolution microscopy. Single-molecule imaging with very dilute samples was an important step,[9] but photo-activation turned the tables.

It is precisely this additional information, the fact of "being sure" to deal with single molecules, which improves the ability to locate the emitters, compared to the limit d indicated by Abbe's law, by a factor equal to $\sqrt{N}$, where N is the number of photons emitted by that single molecule

[9] G. Chirico, A. Diaspro et al., *Single Molecules Studies By Means Of The Two-Photon Fluorescence Distribution*, "Microscopy Research and Technique," 55:359–364, 2001.

already distinctly recognized with others whose position we want to determine. Collecting 10,000 photons, a situation entirely feasible, the accuracy of localization can be attested around 2 nm, considering 200 nm the binding limit of diffraction.

The scheme to obtain super-resolution involves a series of steps. The most critical step concerns the preparation of the sample: it is crucial to mark what we want to visualize with photo-activated fluorescent molecules. These molecules will be activated in subsequent steps, observed, and finally switched off. The activation is carried out with a low light intensity to favour the generation of a few scattered activated molecules with a high probability of being, molecule by molecule, at distances higher than the diffraction limit. This makes them distinguishable. Their excitation produces a fluorescence signal that clearly comes from a single molecule located at some point in the sample. This molecule will put up N photons before finally switching off. When all the photoactivated molecules are turned off, new ones are photoactivated. At this point, the groups of individual fluorescent molecules photoactivated from time to time are mapped with a precision that improves the diffraction limit by a factor $\sqrt{N}$. The procedure is repeated in rapid succession until the image is formed, until the pointillist picture takes shape. Of course, we may have to acquire tens of thousands of image fields and locate millions of positions before we can obtain a super-resolved image at the nanoscale. This solution for optical super-resolution is called PALM, *PhotoActivated Localization Microscopy*, and is based on the localization of photoactivated fluorescent molecules (Fig. 8.7). If marking or photoactivation are not sufficient, or the photons available to localize are few or, worse, individual molecules move from their mutual positions, between localization rounds, we may have problems or limited improvement in spatial

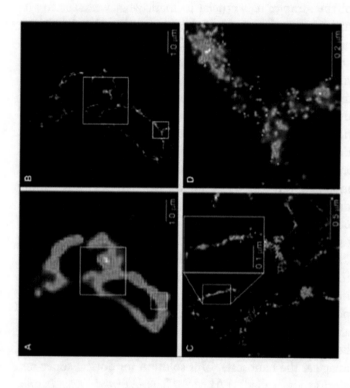

Fig. 8.7 The ability to precisely localize single molecules allows biological molecules to be visualized in increasing detail. This is a thin section of a COS-7 cell...

resolution, which in turn is a close relative of our ability to localize (Fig. 8.8).[10] However, the most interesting question that arises is this: when do I stop repeating the process? When is the image to be considered final?

The key to the answer lies in having precisely formulated the biological question that prompted you to use this method of super-resolution. You certainly can't think of stopping when you recognize what you wanted to see because it makes me say "but you already knew that!". The criterion could be Nyquist's or Shannon's. If anything, you stop when you have reached a density of image points such that the minimum distances between one location and another are about half of the resolution you needed to answer the question you asked. Remember that Nyquist and Shannon have indicated a simple, basic criterion: if you want to reconstruct a signal, in space or time, with a resolution equal to p, you must collect data at least every $p/2$. No less, no more… this is enough. The key to the method remains the "certainty" of the origin of the photons that reach the sensor each time. If the sample is thick? If we were treating an aggregate of tumoral cells? In this case we could, for example, use the three-dimensional method of the light sheet to select portions of the volume occupied by the sample and within each sheet apply the localization of individual molecules.[11] The nanoscale super-resolution scheme can then be realized in a sample a few hundred micrometers thick, thus being able to study the behavior of single molecules in a context such as cell aggregation and in perspective in tissues and organs (Fig. 8.9).

---

[10] H. Deschout, A. Diaspro et al., Precisely *and accurately localizing single emitters in fluorescence microscopy*, "Nature Methods," 11(3):253–66, 2014.

[11] F. Cella Zanacchi, A. Diaspro et al., *Live-cell 3D super-resolution imaging in thick biological samples*, "Nature Methods," 8:1047–1049, 2011.

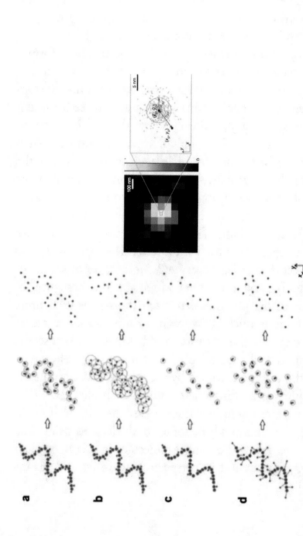

Fig. 8.8 The localization of single molecules, if done correctly, allows us to obtain a faithful map with respect to reality (**a**). If the number of photons emitted by the single molecules is low (**b**), the marking is insufficient even with excellent localization (**c**) or the single molecules move (**d**) the map will not be faithful. The first column represents the real object, the second the location of its individual molecules and the third shows the map-point scattering. The localization accuracy depends on the number of photons emitted by the single molecules

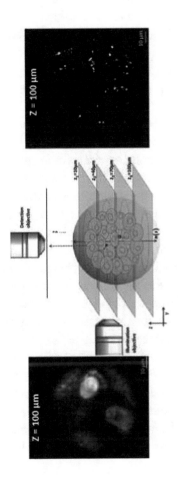

Fig. 8.9 By putting together the light sheet with single molecule localization it is possible to achieve super-resolution on thick samples such as the tumor spheroid shown. On the left is an illuminated section from the leaflet and on the right is the identification of single molecules

Stefan W. Hell, during his doctoral period in Finland, had an interesting intuition, which would prove to be Nobel worthy almost twenty years later, to address the challenge of super-resolution. He thought of using the process of stimulated emission, which harks back to Einstein's theoretical elaboration[12] and the realization of lasers.

When a fluorescent molecule is in an excited state it "lights up", emitting photons in a certain energy region within what is called the emission spectrum. Typically, the fluorescence is observed under the microscope at the wavelength that corresponds to the maximum probability of the emission event so you can also exploit the fluorescence emitted at other wavelengths for the formation of a multicolour image. When the molecule is in the excited state, if instead of letting the fluorescence take the path that is most congenial to it, we force the molecule to a stimulated emission in a permitted region, then we can eventually annihilate the green fluorescence through the emission of red photons. To do this we can use a second illumination beam, where the first is that of classical fluorescence excitation. The experiment involves exciting green fluorescence with blue light to induce a stimulated emission in the red using a source in the red. This type of microscopy is called STED, *STimulated Emission Depletion*. What happens is that as you increase the power of the second beam of light, the green fluorescence decreases. This is the pioneering experiment of Stefan W. Hell applied to image formation[13] that will see its ultimate success in a historic work

---

[12] A. Einstein, *Strahlungs-emission und -absorption nach der Quantentheorie*, 'Verhandlungen der Deutschen Physikalischen Gesellschaft', 18:318–323, 1916.

[13] S.W. Hell, J. Wichman, *Breaking the diffraction resolution limit by stimu-lated emission: stimulated-emission-depletion-microscopy*, "Optics Letters," 19:780–782, 1994.

which appeared in the prestigious journal of the American Academy of Sciences.[14] But if I trigger fluorescence and then erase it, isn't that a contradiction? How can I form an image? The secret, all technological, lies in the fact that we are able to shape the illumination as we want, for example as a doughnut. If I make the second beam of light, the one that serves to depopulate the excited states through the stimulated emission, like a doughnut, it means that at the center of the light beam I preserve the fluorescence that I have triggered and if I "squeeze" as much as possible the hole of the doughnut I obtain a finer and finer spatial selection of the region from which I collect information. Forming the image point by point, it is as if I used a brush with an increasingly fine tip to make my pointillist painting (Fig. 8.10). The spatial resolution improves significantly over Abbe's law because, in this case, d scales by a quantity $\xi = \sqrt{1+(I/Isat)}$ where I is the intensity of the donut illumination beam and I_{sat} is the intensity of that beam which depopulates 50% of the excited states for that specific molecule at a certain wavelength.

The power of the doughnut light beam is the cross and delight of this method. If the fluorescent molecule to be visualized had a very low I_{sat}, you could use a "laser" bought in a stationery store to perform the STED or obtain a high I/I_{sat} ratio. It is in this way that, as a first approximation, you can control the spatial resolution.

On the other hand, when one improves spatial resolution by increasing I, say I/I_{sat}, one also increases the risk of producing an undesired absorption. The increase of I has the potential double drawback of introducing undesired fluorescence and a fast photobleaching process. The

[14]T.A. Klar, S.W. Hell et al., *Fluorescence microscopy with diffraction resolution barrier broken by stimulated emission*, "PNAS", 97: 8206–8210, 2000.

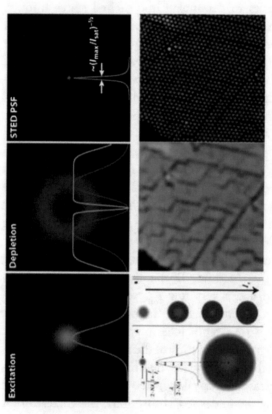

Fig. 8.10 The pointillist painter's brush defined by excitation (*excitation*) is "sharpened" (STED PSF) using a donut beam for stimulated e-mission (*depletion*). The fluorescence used to form the image comes from the center of the donut where the depletion beam has its zero. The more intense the depletion beam, the sharper the fluorescence profile. Fluorescence microscopy of 40 nm fluorescent spheres does not allow them to be spatially resolved. Switching on the depletion beam allows, without the need to do any calculation, to obtain instantaneously the improvement of the spatial resolution

main criticism of the STED approach, since its first public appearance, was related to the aforementioned aspects.
The risk in reality is not so high, and it is worth it considering that the simple switching on of a second source
of illumination produces an immediate improvement of
the spatial resolution and that this is a variable dependent
on the intensity given to the system to realize the depop-
ulation of the excited states. From the excited states, the
fluorescent molecule emitting photons returns to the
base state. Of course, you can eliminate some of them by
taking them on the path of stimulated emission. If the
intensity of the doughnut gets limited, for whatever reason, you can imagine that some green photons manage
to pass "through" the edges. That fluorescence, which you
don't want, however, has one more chance to decay to the
ground state than the fluorescence you observe coming
from the center of the doughnut. The reasoning is a bit
convoluted, but we can simplify by saying that the lifetime
of the unwanted fluorescence is shorter than that of the
unperturbed one.

This observation allows us to use the arrival time of
photons at the sensor to know whether they come from
the center of the doughnut or the periphery. We can
accept or discard photons for our image by having an
extra knob to select the spatial resolution (Fig. 8.11), or
to classify them by arrival time to improve the spatial resolution for the same $I/I_{sat.}$ The method is called SPLIT,
Separation of Photons by LIfetime Tuning,[15] and we can say
that it descends from the idea of making a STED-type

[15] L. Lanzanò et al., *Encoding and decoding spatio-temporal information for super-resolution microscopy*, "Nature communications," 6:6701, 2015.

Super eyes to see light signals129

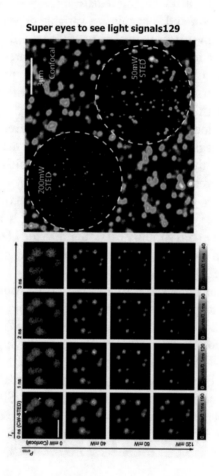

Fig. 8.11 The time factor can improve the use of STED microscopy allowing to obtain, with the same depletion intensity, a better resolution. The image shows how from the confocal scheme where the intensity of depletion is zero you can improve the resolution passing first to 50 mW and then to 200 mW. *Credit* Giuseppe Vicidomini and Diaspro Lab, IIT

microscope with arrival time selection, called "Gated" STED.[16] The need to bring super-resolution STED into organs and tissues has been met by making several STED schemes with two-photon excitation. Amongst them, and one of the most promising approaches, uses an illumination beam and donut of the same colour to excite and depopulate due to different pulse durations.[17] STED today represents the most flexible and effective way to observe the living at the nanoscale (Fig. 8.12).[18]

So, the two "Nobel methods" differ conceptually and practically in the way they make our eyes become super eyes. In the first case the stochasticity of the events on which the localization of the single molecules is based is in force, while the second one is based on the stochasticity of the events on which the localization of the single molecules is based knows about the precise determination of the region within which to make the event occur. Events that are fundamental for obtaining the super-resolution in fluorescence microscopy and that clearly reiterate the role of the probe, of the preparation of the sample rather than of the chosen optical architecture. One of the most important points lies not only in the ability to reveal what previously could not be grasped, possibly only imagined, but also in the possibility of being quantitative. Of being able to count the molecules involved in the life of the biological cell.

---

[16] G. Vicidomini et al. *Sharper low-power STED nanoscopy by time gating*, "Nature Methods," 8:571–573, 2011.

[17] P. Bianchini, A. Diaspro et al., Single-wavelength two-photon excitation-stimulated emission depletion (SW2PE-STED) superresolution imaging, "PNAS", 109(17):6390–6393, 2012.

[18] G. Vicidomini, P. Bianchini, A. Diaspro, *STED super-resolved microscopy*, "Nature methods," 15(3):173–182, 2018.

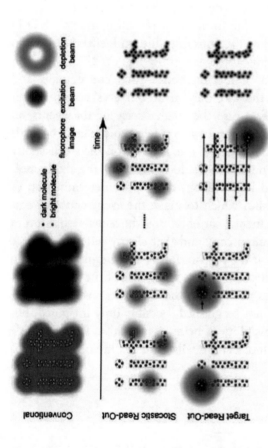

Fig. 8.12 The 2014 Nobel super-resolution microscopy offers two possible ways forward united by the ability to control the fluorescence phenomenon. The appreciable achievement of recognizing "iit" is made even more revolutionary by the fact that it begins to be possible to count individual emitters, to be truly quantitative. *Credit* Giuseppe Vicidomini and Diaspro Lab, IIT

9

Label-Free

*It's 126 miles to Chicago. We have a full tank of gas, half a pack of
cigarettes, it's dark, and we're both wearing sunglasses. Go!*
Dan Aykroyd to John Belushi, in *Blues Brothers*, 1980

DNA in eukaryotes is mainly organized inside the nucleus
of the biological cell compacting itself in different levels of
organization. It is incredible how that filament, organized
in a double helix about 2 nm wide, compacts its 2 linear
meters in a few millionths of a meter.[1]

It is even more interesting how the information needed
for differentiation, for the life of the cell, is released where
and when it is needed. In the transformations of the cell,
the DNA organizes its molecules, its base-pairs super-or-
ganizing itself in complexes scaled from tens of nanometers to hundreds up to thousands in the 46 chromosomes
of the man that contain about 5 cm each (Fig. 9.1).

---

[1] K. E. Van Holde, *Chromatin, Springer Series,* "Molecular Biology," 497, 1989.

© Casa Editrice Ulrico Hoepli S.p.A. 2022
A. Diaspro, *Expedition into the Nanoworld*,
https://doi.org/10.1007/978-3-030-94472-8_9

The key biological macromolecule in the processes that regulate DNA compaction is chromatin. Chromatin, which since 1879 has been named after the German biologist Walther Flemming (1843–1905), is composed of histones, basic proteins, and DNA, or 147 base pairs of DNA, that wrap around 8 histones, the histone octamer, to form the nucleosome, its basic unit.

The nucleosomes organize themselves six at a time in the turns of a helix trajectory that will repeatedly compact inside the nucleus. It is precisely the organization in the nucleus, the interaction of the various macromolecular complexes with each other, that determines the life of a cell, of sets of cells, organs and tissues, of human beings and also their mutual interactions down to the meter scale.

Using a classical terminology, which, like all terminologies related to the living being, is changing due to the details with which, little by little, we learn to know the mechanisms of life, we distinguish two types of chromatin. The euchromatin, which is less intensely "marked" and represents an organization at a relatively low level of compactness. And the heterochromatin, which is more intensely "marked" and represents an organization at a relatively low level of compactness.

Marked represents more condensed areas, a higher level of compactness of the information written along the DNA. The genetic meaning is different, at least for the current state of knowledge. In heterochromatin DNA is considered inactive and is not transcribed into RNA. Therefore, the information contained in DNA does not participate in the workings of the cellular nanomachine leading us to consider heterochromatin as the inert state.

Euchromatin, on the other hand, represents a state in which DNA is "active" and makes its information available, at a certain moment in the life of the cell, in a specific way (Fig. 9.2). The organization of chromatin, either in

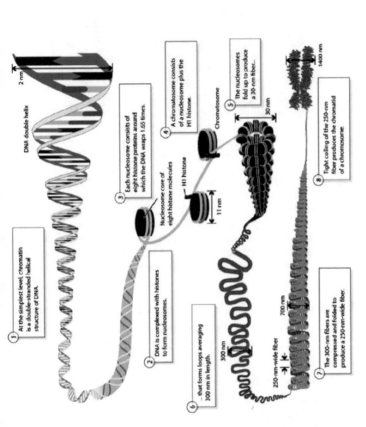

Fig. 9.1 Hypothesis of chromatin organization. Observing levels of organization at a spatial scale ranging from about 2–1400 nm with the same light microscope is part of a possible challenge

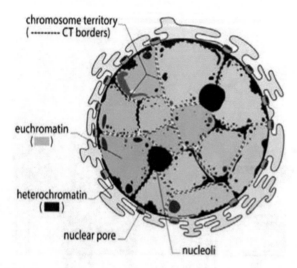

Fig. 9.2 The chromatin that compacts and decomposes in the nucleus, allowing the living being to release or not the information encoded in the DNA, has an irresistible beam. The system has such a high density of biological molecules that even small fluorescent markers could change its arrangements. The prospect is to use *label-free* methods to follow processes and ideally be able to make, for example, pre-neoplastic diagnoses

isolation or in nuclei, is so important that over the years several models have been imagined and hypothesized, many of them developed based on electron microscope observations. Unveiled chromatin is a flexibly joined chain of nucleosomes that exists within the nucleus of eukaryotic cells, described as "a necklace of pearls" by the studies of Sir Aaron Klug (1926–2018), Lithuanian biophysicist naturalized British, Nobel Prize in Chemistry in 1982 "for his development of crystallographic electron microscopy and for elucidating the structure of biologically important nucleic acid-protein complexes".

It is a fact that chromatin organization strengthens DNA in those mechanically stressful processes such as mitosis, helps protect DNA from damage, and is

Fig. 9.3 The cover dedicated to the new chromatin organization hypothesis

important in controlling gene expression and DNA replication.[2] On July 28, 2017, a beautiful cover story in "Science" (Fig. 9.3) illustrated the possibility of a new organizational model of chaotic groups of cat-of this biopolymer on the order of 5–24 nm, which the authors of the research considered to be more functional for the tasks it had to perform.[3] Among the most interesting aspects in studying the organization of the chromatin, there is that which concerns the influence of external factors, from the environment to the climate, from your last anger to your first kiss. The influence of those factors that make the DNA clump change, without changing the sequence of base pairs, and consequently the genetic apparatus on which the British biologist Conrad Hal

[2] G. Li, D. Reinberg, *Chromatin higher-order structures and gene regulation,* "Current opinion in genetics and development," 21(2):175–186, 2011.

[3] H. D. Ou et al., ChromEMT: *Visualizing 3D chromatin structure and compaction in interphase and mitotic cells,* "Science," 357(6349):eaag0025, 2017.

Waddington (1905–1975) focused on when he coined the term "epigenetics" in 1942.[4] Waddington thus intended to define "the branch of biology that studies the causal interactions between genes and their cellular product and brings into being the phenotype". An idea of which conceptual elements can be found in Aristotle (384–322 B.C.), who believed in the development of individual organic forms from the "unformed" or epigenesis.

In short, it has been a fascinating and hot topic for over two thousand years. The existence of so many models and hypotheses on chromatin organization seems to indicate that we still need to refine our tools, our way of designing experiments and reading data.

The tools we have at our disposal are incredibly powerful, giving our eyes super vision and allowing us to tackle detail at the scale we need most. The computing and data storage power we have at our disposal allows us to collect unimaginable amounts of data.

The markings on our samples are precise and produce many messages when the light of the rainbow intercepts them, and our piece of glass picks up the signals. It is precisely here that we plan to intervene. The reason is simple after all and concerns crowding, molecular crowding. It is well known that the density of information is very high and the level of compaction of chromatin in the nucleus, and not only of chromatin if we think of the multiple activities of the cell with its macromolecules, is also very high.

---

[4] C. H. Waddington, *The Strategy of the Genes*, Taylor & Francis, 2015 [Allen & Unwin, 1957].

Well, imagine that you take public transport every morning to go to work or school. Even though you take it during rush hour and it is always crowded, when your stop is approaching you know how to arrange yourself among the people, how to organize yourself to get off at the right time. Now, instead, try to imagine what would happen if one morning someone put on you, like all the other passengers, a backpack of similar size to yours. Would you be able to quickly reorganize to gain an exit at the right time? Would people get organized like they do every other day?

After all, when you label chromatin or other cellular components with a fluorescent molecule, you are doing just that someone else did on the vehicle that takes you to work or school. Simply, after labelling, you have increased the "molecular" density, imagining that this does not change anything with respect to the spatial organization of the molecules or their function. This may be so, but don't we want to try and see if we can get the same information with just light and a lens from the living without the need for marking? Observing without the need of labels being placed in the right place and located. We are a bit like acrobats, without a net, suspended between the light messages we intercept trying to understand who is sending them and why.

There are several ways to get images without the need for marking, from Zernike phase contrast to autofluorescence to second harmonic signal, we know that.

If we consider a particular property of light, polarization, we can enter the world of Mueller's matrix microscopy, a rich light paraphernalia that makes the eyes see what is transparent, what they cannot observe, without "planting flags". The formalism is owed to Swiss MIT Boston physicist Hans Mueller (1900–1965), who used an interesting description of polarized light suggested by Baronet Sir George Gabriel Stokes (1819–1903), an

Anglo-Irish physicist and mathematician, in 1852. Stokes made incredible contributions to physics and mathematics in various fields, and it is no coincidence that, because of his interest in the interactions between the light of the rainbow and a piece of glass, we have already met him in the description of the phenomenon of fluorescence.

But now we are interested in an interesting property of light, which we have already mentioned, which is polarization.

I like to associate this phenomenon with the coffee machine that you find everywhere, including research laboratories. In research laboratories it becomes one of the best places to give birth to new ideas, insights, thanks to that mixture of relaxation and sharing the queue with researchers who study things that are often different from our own. But ideas, new ideas, are universal. Some may specialize in them, but they are born as a universal need and a scientific need.

The researcher, who does not avoid the "method", once has his or her own idea, spends his days trying to demonstrate it. But what does the coffee machine have to do with polarization? Imagine that you want to be original and instead of inserting the coin you try to throw it towards the machine. The queue behind you would probably be stretched to its limits waiting for your attempt to hit the slot with the coin toss in order to get your coffee. The probability of the event is much lower than when along the French highways you "threw" or tossed change into the basket put there to pay the toll. With the coffee machine you have to make two events: center the slot, and center it with the coin oriented so that it enters exactly in the slot. When the coin enters that slot, it enters in a "polarized" way, and if you try to make it enter by "pushing" it oriented at ninety degrees to the slot, I'd say no way. On the other hand, when you throw coins into the basket you can do it in a "non-polarized" way.

For light, for the electromagnetic field that we associate with light, things are similar.

In physics, the polarization of the electromagnetic radiation of light is a characteristic of electromagnetic waves and indicates the direction of the oscillation of the electric field vector during the propagation of the wave in space and time.[5] The magnetic field, as usual, will be polarized along the direction orthogonal to that of the electric field and to the direction of propagation, which we have already associated several times with the z axis of propagation of illumination in our microscope.

The expression that makes light a polarized light was introduced in 1808 by the French physicist, mathematician, and engineer Étienne-Louis Malus (1775–1812) while conducting experiments on light reflection.

The concept of polarization refers to the mode of oscillation of the electromagnetic wave associated with light radiation. If it always oscillates in a vertical plane, we say that the wave is polarized vertically. Consequently, if it always oscillates in a horizontal plane we will say that the wave is horizontally polarized. If the electric field rotates as it propagates around the direction of propagation, with always identical amplitude, then we will say that the wave is circularly polarized (Fig. 9.4). Any other direction of polarization can be obtained as a superposition of a horizontally polarized wave and a vertically polarized wave. In spacetime there is no conserved mode of oscillation, or different sources of light propagate with different oscillations, and what we generically collect is unpolarized

---

[5] A. Bettini, *A Course in Classical Physics, Vol. 4—Waves and Light. Springer,* 2016.

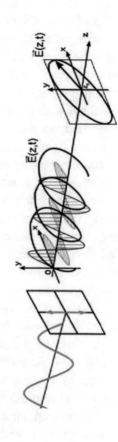

Fig. 9.4 Light is an oscillating electromagnetic field and this defines its polarization in space and time

light. By non-polarized light we mean all the light that reaches our eyes or a sensor without being "discriminated" in terms of wave orientation or polarization state. The famous polaroid glasses exploit the fact that when light meets a surface and is reflected, it acquires a prevailing polarization state; therefore, if those glasses are made to let the orthogonal polarization state pass, a bit like in the case of the coffee machine when you try to insert the coin at ninety degrees with respect to the orientation of the slot, you can attenuate the sunlight that had arrived to the reflecting objects in a non-polarized way.

In 1852 Stokes defined a set of parameters to describe the polarization state of the electromagnetic field associated with light.[6]

These parameters, which can be used at the input and output of our linear space invariant system, the optical microscope, can be used to form images. Stokes suggested classifying light using four parameters which he put into a vector of four elements, the Stokes vector. The elements are *I*, *Q*, *U* and *V*. Considering *I* all light in any state of polarization, *Q, U, V* are respectively the normalized percentage between +1 and −1 of light polarized horizontally (+1) or vertically (−1), at +45 (+1) and −45 (−1) degrees in the plane of polarization and with rotation to the right (+1) or left (−1) for circular states. If our radiation were all circularly polarized to the left, the Stokes vector would represent it as [1, 0, 0, 1] with the understanding that *I2 Q2 + U2 + V2.*

Since we can control very well the polarization states of light, the illumination, we can describe with a Stokes

---

[6] G. C. Stokes, *On the composition and resolution of streams of polarized light from different sources*, "Transactions of the Cambridge Philosophical Society," 9:399–416, 1852.

vector $_{[Sin]}$ the polarization that reaches our sample. When we detect the signal with our sensors, we are able to efficiently select the states of polarization, that is to describe the radiation that reaches the sensor again with a Stokes vector, $[S_{out}]$ originated from the interaction of $[S_{in}]$ with the specimen.

Now, if the Stokes vector arriving at the sensor is different from the illumination vector, the responsibility for the change lies with what the polarized light encountered along the way. How could we describe the transformation of $[S_{in}]$ into $[S_{out}]$? What a question, there are two vectors of four elements, so a 4×4 matrix manages to do that and describes, through its values, how the objects encountered along the optical path change the polarization state of the light. This is the Mueller matrix, $[M]$.[7] We can therefore write that $[S_{out}] = [M][S_{in}]$. Each optical element of the system can be described by a Mueller matrix and the set of elements, including the sample, encountered by the illumination vector can be considered through its own Mueller matrix which will enter a series of multiplications between matrices. What matters to us now is not the power of the Mueller matrix in designing the instrument, but in describing the behavior of our observed system through its sixteen elements, all of which are measurable, i.e., obtainable from the measurement of the parameters of $[S_{out}]$ known those of the investigating light $[S_{in}]$ (Fig. 9.5).

Can the interaction of the polarized light with the specimen provide us with structural information and put us in a position to obtain a contrast with which to make an image, point by point?

[7] H. Mueller, The Foundation of Optics, "Journal of the Optical Society of America," 38, 661–663, 1948.

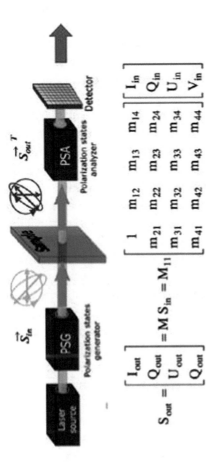

Fig. 9.5 The scheme is simple: you control the polarization of the illumination and the polarization of the light at the sensor. We attribute the variations to the sample. A vector of 4 elements describes the input and output polarization, and a matrix of 16 elements describes the effect of the sample. Thus, 16 different images can be produced. *Credits* Aymeric Le Gratiet, Diaspro Lab, iit

From a phenomenological point of view, it has been shown that the sixteen images that can be made, point by point, by mapping the elements of the Mueller matrix are representative of whether, for example, a healthy rather than diseased tissue is being analyzed, with reasonable reproducibility.[8]

Let's try to understand what information we can get about the biological molecules that are in the cell using images made with the different elements of the Mueller matrix.

The first element of the Mueller matrix, [*M00*], has to do with all the light we can collect regardless of the state of polarization. The second element of the matrix, on the same line, [*M01*], can be shown to have something to do with the phenomenon of birefringence, i.e. the ability of a material, a crystal, to decompose a ray of light into two rays depending on the polarization of the light. Thus, other elements can be explained by considering to what mixture of elements of the input and output Stokes vector they owe their value.

In collecting the behavior of the sample with respect to the polarization of light, it is possible to add, for example by varying the numerical aperture of the collecting lens or by angling the sensor, additional information regarding the way in which the interaction between matter and light is scattered, diffused. This phenomenon is known as *scattering*.

Thus, a class of phenomena of interaction of radiation with matter whereby waves or particles change their trajectory due to in-counter with other particles or waves. One

---

[8] Le Gratiet, A.; Mohebi, A.; Callegari, F.; Bianchini, P.; Diaspro, A. Review on Complete Mueller Matrix Optical Scanning Microscopy Imaging. Appl. Sci. 2021, 11, 1632.

can understand many properties of the objects we intercept with the light beam by collecting their scattering at different angles. For example, the [*M00*] element of the Mueller matrix accounts for the size of the intercepted objects along the propagation direction at the zero scattering angle, while if detected at ninety degrees it provides information about their shape. In the panorama of the use of polarized light we are interested, for the type of biological question about the organization of DNA that we have, in a particular element of the Mueller matrix, the element [*M03*], which "says" if and how much our preparation lets pass or scatters light polarized circularly to the right compared to that polarized circularly to the left.

After all, you may find it normal that light arriving by rotating to the right or left may have a different interaction with an object organized as a right or left-handed helix and having a certain radius and pitch of the helix itself. The ability of some molecules to interact differently with the incident circularly polarized light when the rotation is to the right or left is called optical activity. These molecules are known as chiral molecules and have the property of being non-overlapping with their own mirror image. Hands are an example of a chiral "object" because the right and left hands are mirror images of each other, but they cannot be superimposed. The pair of chains that makes up the DNA strand, wrapped in a double helix is chiral in the distinction between the right-handed and left-handed helices. The two phenomena traditionally associated with optical activity are *Optical Rotatory Dispersion* (ORD) and *Circular Dichroism* (CD). The ORD allows evaluation of the rotation of the plane of polarization of light as a function of wavelength of light due to the difference in the propagation velocity of the two circular polarizations through the material. While the

CD provides a measure of the difference in the absorption cross section for the two incident circular polarizations.[9]

But didn't we say that under rainbow light DNA and proteins are transparent, they don't absorb? True. During measurements outside the absorption band of DNA and proteins, typically between 280 and 260 nm, an unexpected signal was measured. The signal was due to differential scattering of light when, although outside the absorption bands, it encountered the various levels of structure in helices and superhelices of DNA and proteins being polarized once in a circular manner with right-handed rotation and another time left-handed. Initially, the understanding of this effect was used to correct possible artifacts, but very soon the enormous potential was realized.[10] It was differential scattering of light circularly polarized to the right and left. By changing the wavelength of illumination out of the absorption band it was possible to demonstrate that DNA and proteins when they organized themselves with different steps and rays in helical trajectories produced a signal attributable to low compaction. In 1980 Carlos Bustamante, a Peruvian scientist born in 1951 and an American biophysicist, a pupil of Ignacio Tinoco Jr. (1930–2016), is one of the fathers of studies on the packing of RNA and ribonucleic acids. He published his doctoral thesis in the laboratory of the young Marcos Maestre, "fresh" from his doctorate received at Yale, Berkeley[11] on the study of chiral molecules with differential scattering of polarized light in circular states—CIDS,

---

[9] I. Tinoco, M. F. Maestre, C. Bustamante, *Circular dichroism in samples which scatter light*, "Trends in Biochemical Sciences," 8(2):41–44, 1983.

[10] B. P. Dorman et al., *Experimental differential light-scattering correction to the circular dichroism of bacteriophage T2*, PNAS, 70(1):255–259, 1973.

[11] C. Bustamante, *Circular intensity differential scattering of chiral mole-cules*, Ph.D. Thesis California University, 1980.

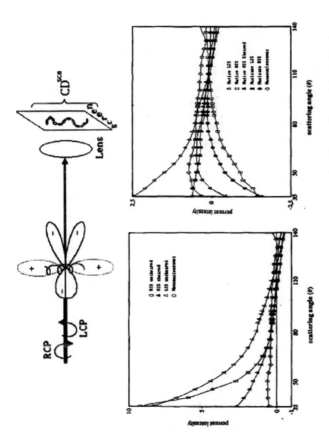

Fig. 9.6 The fourth element in the first row is sensitive to pitch and radius of possible helical or superhelical organizations. The different trends in the graphs depend on having analyzed differentially organized chromatin in solution in a controlled manner (Diaspro et al., *op. cit.*, 1991)

Circular Intensity Differential Scattering. In the following years several investigators on CIDS started to be made public by the scientific community.[12]

It will be the interest in chromatin, the desire to keep faith with the promise made to my grandmother Anna, that will lead me to engage in a theoretical and experimental "assault" on the use of CIDS to study the levels of organization of chromatin.[13] The meeting, years later, with Ignacio Tinoco Jr. and Carlos Bustamante, will convince me, on the one hand of the work still to be done and, on the other, of the high potential of CIDS to unveil the secrets of that chromatin organization and of its regulation of "the good and bad times" in our lives. In 1991 it will be possible to demonstrate how with red light the 632.8 nm line of the He-Ne laser can be used to obtain CIDS measurements from chromatin compacted differently in solution (Fig. 9.6).[14] Following that same line of research it was possible to use CIDS to distinguish different bacterial cell populations by small changes in DNA organization.[15]

Here, CIDS is an excellent candidate to study chromatin compaction in cellular systems without the need to use

---

[12] K. S. Wells et al., *An analysis of circular intensity differential scattering measurements: Studies on the sperm cell of Eledone cirrhosa*, Biopolymers, 25:2043–2064, 1986.

[13] A. Diaspro C Nicolini, *Circular Intensity Differential Scattering and Chromatin-DNA Structure A Combined Theoretical Approach*, "Cell Biochemistry and Biophysics", 10(1):45–60, 1987.

[14] A. Diaspro et al., *Polarized Light Scattering of Nucleosomes and Polynucleosomes-In Situ and In Vitro Studies*. "IEEE Transactions of Biomedical Engineering," 38(7):670–678, 1991.

[15] A. Diaspro et al., *Polarized Light Scattering: A biophysical method for studying bacterial cells*, "IEEE Transactions on Biomedical Engineering," 42(10):10,381,043, 1995.

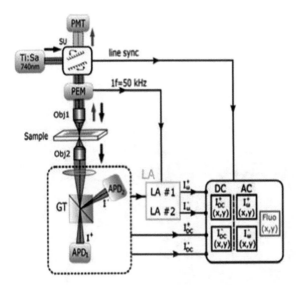

Fig. 9.7 Schematic of the multimodal microscope for fluorescence and CIDS (Le Gratiet et al., op.cit., 2019)

fluorescence or other markings, producing images by mapping the [*M03*] element of the Mueller matrix.[16]

The fact is that we are really tightrope walkers without a net, we are collecting photons from crowded environments, we are trying to classify them by their possible origin and to build images of something that our eyes cannot see and our reasoning cannot imagine.

To answer the scientific question that was posed, let's go back to our piece of glass and the rainbow. Moving towards *label-free* microscopy does not preclude using everything we have learned to better understand what "netless" photons can tell us. We can use everything that

---

[16] W. Mickols, M. F. Maestre, *Scanning differential polarization microscope: Its use to image linear and circular differential scattering*, "Review of Scientific Instruments," 59:867, 1988.

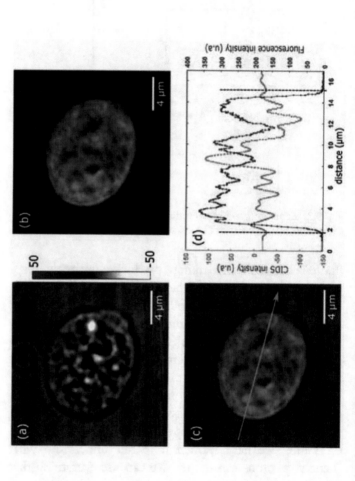

Fig. 9.8 Combined CIDS (a) and fluorescent (b) image (c) and associated line profile (d) (*a, Gattica et al., reproduced, 2018)*

has made our eyes super eyes and exploit everything we know how to do with fluorescent molecules to place alongside the *label-free* images we construct using the Mueller matrix method those images we already know how to use and interpret. So, for a while we will be acrobats, tightrope walkers, tied up with a beautiful mountaineering harness. At the right moment we will unhook the bindings, we will get rid of the harness, of the fluorescent molecules, and we will be really without net and *label-free*.

Let's start now. We use two-photon instead of one to excite the fluorescent molecules that generally mark chromatin, and we use that same deep red beam, whose polarization we control by modeling the Stokes vector of illumination, to derive a point-by-point map of CIDS. We build an optical microscope that is the sum of two microscopes and that uses the same piece of glass and the same light as the rainbow to make our eyes see something that "*seems impossible until it is done*" (Fig. 9.7), as Nelson Mandela (1918–2013) said.

The organization of DNA in the superattracted strand that is chromatin is thus revealed in the nucleus of the cell so that it can be deciphered,[17,18] (Fig. 9.8).

They are "cards to be deciphered" (Alan Parsons, *The turn of the friendly card*, 1980) in order to follow the changes of compaction in living cells, in aggregates of cells, in organs, in tissues and in living organisms.

---

[17] A. Le Gratiet, A. Diaspro et al., *Circular intensity differential scattering (CIDS) scanning microscopy to image chromatin-DNA nuclear organization*, "OSA Continuum," 1(3):1068–1078, 2018.

[18] A. Le Gratiet, L. Lanzano, A. Bendandi, R. Marongiu, P. Bianchini, C. J. R. Sheppard, A. Diaspro, "*Phasor approach of Mueller matrix optical scanning microscopy for biological tissue imaging*," Biophysical Journal, 120(15): 3112–3125,2021.

10

The Future of Microscopy

Liquid Images Between Multi-messenger Microscopy and Artificial Intelligence

It was winter in Belleville and there were five characters. Six if you counted the sheet of ice. Seven, indeed, with the dog that had accompanied Piccolo to the baker's.
Daniel Pennac, *The Fairy Gunmother*, 2003
If you can't bring good news, then don't bring any.
Bob Dylan, The wicked messenger, in "John Wesley Harding," 1967.
You've spent a long time waiting
For a perfect yesterday,
Now fill your heart with celebration
For that's love's way...
Premiata Forneria Marconi,
Celebration, in "Photos of Ghosts"
lyrics by M.Pagani, P. Sinfield,
music by F.Mussida,
F. Premoli, 1974.

© Casa Editrice Ulrico Hoepli S.p.A. 2022
A. Diaspro, *Expedition into the Nanoworld*,
https://doi.org/10.1007/978-3-030-94472-8_10

My granddaughter Irene, Claudia's and Francesco's daughter, who is a very fresh five years old since "when she was little", as she explains to me today, was attracted by those colourful pictures she saw around the house.

Those figures that with mum Claudia have become a "Pop" collection. One of them, green as the hills of Ireland, was dedicated to her before she arrived, Ireland or better "Irene land!" How many things can be done with those images that the eyes cannot see, revealed by a piece of glass and a rainbow.

Irene asks me about the future of microscopy and in the meantime, she plays with BLIPS pulling out from a secret hiding place a piece of banana that she had left "spoiled" to observe those small bacterial communities that make it "go bad".

The future of microscopy? I have no doubt. The future of microscopy lies in "liquitopy"[1] and the construction of a multi-messenger microscope that I call LDM, *Light Diversity Microscope*. "Light diversity" is almost untranslatable, but it simply wants to emphasize the different way I treat light to produce and understand the images that the optical microscope forms. I can use geometric optics to track light rays, electromagnetic waves to define polarization states, or photons to explain localization enhancement in single molecule microscopy. What diversity in the treatment of light!

Combining Different Microscopy Methods to Expand the Amount Of Information Is Part Of The Current Debate And Includes Just "Liquitopy".[2]

"Liquitopy" stands for *LIQUId Tunable microscOPY*. "Tunable" stands for "tunable" and accounts for our ability

---

[1] © A. Diaspro, registered verbal trademark.

[2] R. Won, *The super-resolution debate*, "Nature Photonics," 12:259–261, 2018.

to see the useful spatial resolution for the problem at hand, for our super eyes to pick up variable detail.

"Microscopy" we know by now, you know it well, is the microscopy made with a more or less curved piece of glass and rainbow light.

"Liquid" is the positive reference to Polish sociologist Zygmunt Bauman's (1925–2017) and his idea of a liquid society elaborated in conducting an original analysis of social stratification and the labour movement in that contemporary phase of transition from modern to postmodern culture.[3]

"Liquid" accounts for the fact that you have a "melting pot" of methods using a piece of glass and rainbow light. Why not put them together in a "liquid" way, almost as if you wanted to make a good "minestrone alla Genovese", tasty soup made using vegetables in the Genoa area, that by putting everything in the cauldron, if well done, allows those who eat it to feel the individual flavour, exalted by others, and to receive an incredible new aroma that makes it more palatable. The construction of the multi-messenger microscope, LDM, precisely concerns the liquid sharing of all those methods available today (Fig. 10.1). It's a practical realisation made of modern and original solutions, starting from the illumination module and arriving at the collection of messages that, through photons, reach the sensor or sensors after interaction with the sample that releases them (Fig. 10.2).

"Liquid" means being able to produce a liquid image, an image in addition to all those that can be produced with every single microscopy technique.[4]   An image in

[3] Z. Bauman, *Liquid Modernity*, Laterza, 2011.

[4] A. Diaspro, *Nanoscopy and Multidimensional Optical Fluorescence Micro-copy*, Chapman and Hall/CRC, 2010

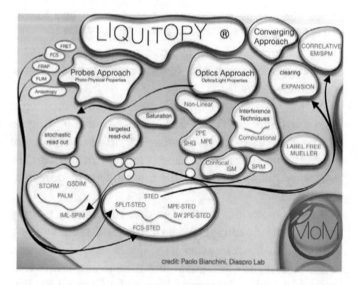

Fig. 10.1 Main methods integrated into a single *liquitopy-based* multi-messenger microscope that we call MOMIX (*Multimodal Optical MIcroscopy Image Correlation Sensing*), © A. Diaspro, registered verbal trademark

which at every point the pointillist painter puts something that reminds us of his work but everything he is able to perceive but without the obsession of having to represent reality, even the reality of measurement, just as it is. Or as he expects it to be. It is a new image, unpredictable for the details it offers, and which overcomes established habits and possible "biases", the usual ways in which the image represents reality. It is not the classic X-ray image in the hands of the expert radiologist who sometimes does not distinguish between a small cyst or a tumor that, if taken at that moment, can be eliminated. It is a new image that can give an intuition which in turn can be verified by studying all the images produced by LDM.

On November 13th, 1984, the morning after Claudia's birth, I was traveling by train to my first conferences as

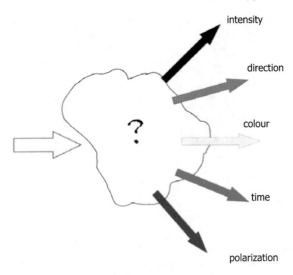

Fig. 10.2 We do not limit the effects of light-matter interaction to the evaluation of light intensity. We use in a classical and original way all the other properties of light that can give us information about the characteristics of the samples we want to study

a selected speaker. I was excited and agitated, we had decided with Teresa that I would not have missed that appointment but would have gone back immediately, to see her and Claudia at the Hospital. It was about 6.30 in the morning, and I would be back shortly after lunch: it seems to me that trains to Milan were even faster than today.

In the train compartment, almost as if in a dream, Vincenzo Tagliasco (1941–2008) appeared, a precursor as well as a pioneer.

Vincenzo Tagliasco, a pioneer at University of Genoa in artificial intelligence, robotics, and the use of images, listened to my story as a new father, congratulated me on my professionalism in not skipping a commitment and offered me a fragrant and tasty tangerine taken from

Fig. 10.3 Constructing, point by point, an additional, liquid image with respect to the conventional ones can allow us to grasp details that had escaped a classical and "habitual" inspection of the image. Klee represents the liquid image and Botticelli the "classical" one

a shopping bag that he held in his right hand. We talked about science and that discipline that is eidology, the science of vision, form and appearance. I think I understood then how important it was not only to produce images and treat them but also to present them, to use them as a synthesis of the creation of quantitative data to make decisions, medical decisions being among the most significant. Shortly afterwards a text came out collecting the images of that congress, edited by Vincenzo Tagliasco himself, more than a university professor for those who had been his pupil and had known him.[5]

[5] V. Tagliasco (ed.), *Eidologia medica: computer graphics in medicine*. Proceedings of the IV Conference Aicographics (Milan), Jackson, 1984.

As we were saying, a new image (Fig. 10.3). After all, from Botticelli's Birth of Venus, which you may or may not like, you don't learn many more things about anatomy; if you then look at a painting by Klee, you may not recognize your own home, but you may find details that will induce you to try to know something more about what you see.

This is not microscopy in the dark but a new paradigm for the microscopy of the future, the future that is already today.

If the collection of information passes through that set of image data provided, for example, by CIDS and fluorescence, or by the super-resolution *label -free* that allows us to count monoatomic layers of graphene,[6] the material of the future, we must use methods of recognition and mathematical description of what we put on our pointillist canvas. The recognition of shapes and objects, their collection in clusters or homogeneous aggregates,[7] the use of associative memory[8] to corrupt the image to produce a new one, allowing you to recognise details previously invisible in the image. Just one of the keys to using the microscope of the future.

Here's where our multi-powered, multi-messenger optical microscope, based on the different way we process and read light signals, needs a brain to untangle itself. Don't think of an automated or robotic system, think of a cognitive, intuitive system. For microscopy, for that piece

---

[6] G. Zanini, A. Diaspro et al., *Label-Free Optical Nanoscopy of Single-Layer Graphene*, "ACS Nano," 13(8): 9673–9681, 2019.

[7] M. Ying Yang, B. Rosenhahn, V. Murino, *Multimodal Scene Understanding: Algorithms, Applications and Deep Learning*, Academic Press, 2019.

[8] A. Diaspro, G.C. Parodi, R. Zunino, *A performance analysis of an associative system for image classification*, "Pattern Recognition Letters", 14(11):861,868, 1993.

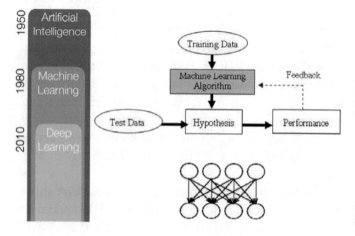

Fig. 10.4 The power of artificial intelligence in its declinations is an indispensable element for the modern microscope

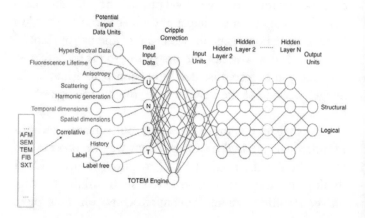

Fig. 10.5 The layers of a dynamic learning approach will benefit from multimodal optical methods, non-optical microscopy methods, and hopefully literature published in at least the last 50 years

of glass and the light of the rainbow, the time has come to meet artificial intelligence. Meet it not because it is

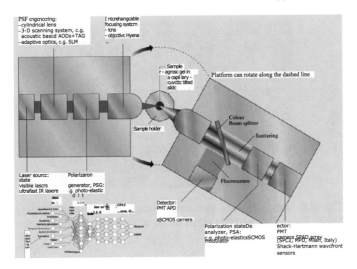

Fig. 10.6 Schematic of the LDM with the integrated artificial intelligence

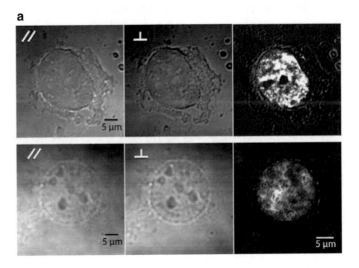

Fig. 10.7 Multimodal imaging of cell nuclei related to DNA organization

b

Fig. 10.8 Are you sure you can count all the wooden blocks? Credit for the idea to Rodolfo Zunino of the University of Genoa

fashionable, not with a joke, but with a path that embeds artificial intelligence from the dawn of the 1950s, that intelligence exhibited by machines rather than humans. In the 1980s *machine learning* algorithms that learn from experience without the need for explicit programming burst into the scene, until the first decade of the 2000s when the algorithms, the programs, modify through experience the criteria with which they decide the parameters used to make decisions (Fig. 10.4).[9]

Here is the artificial intelligence, an additional but decisive character in our story, in the design of MOMIX (Figs. 10.5 and 10.6).

For this intelligent microscopy, super eyes will allow one to see things (Fig. 10.7) that no one has ever seen beyond "the ramparts of Orion and the gates of Tannhäuser" (R. Scott, *Blade Runner*, 1982).

It takes the eyes of a child to see in a woodpile something more, it takes the optical microscope in his new

---

[9]Y. LeCun, Y. Bengio, G. Hinton, *Deep Learning*, "Nature," 521:436–444, 2015.

childhood for that beautiful infinite story between a piece of glass and the rainbow (Fig. 10.8).

See also, Ali Mohebi, Aymeric Le Gratiet, Riccardo Marongiu, Fabio Callegari, Paolo Bianchini, and Alberto Diaspro, "Combined approach using circular intensity differential scattering microscopy under phasor map data analysis," Appl. Opt. 60, 1558–1565 (2021).

Once, when I was six years old […] I managed, with a coloured pencil, to trace my first drawing. My number 1 drawing… When I showed my masterpiece to the adults, I wanted to know if they were afraid of it. They replied, "Why would a hat scare us?" My drawing did not depict a hat, but rather a boa snake digesting an elephant. So, I drew the inside of the snake so the adults could understand. They always need explanations. My drawing number 2 was like this....[10]

disegno numero 1

disegno numero 2

---

[10] Adapted freely from: Antoine de Saint-Exupéry, *The Little Prince*, 1943.

Pop Microscopy

There is a big difference between photography and image: the first one must document, the second one can be creative even if not necessarily truthful. That's why I like to talk about the ethics of photography.
Gianni Berengo Gardin, photographer and photojournalist (1930).

With Claudia we observe how a dewdrop on a leaf lapped by spring sunbeams enhances the detail of the veins and reflects the luminous messages of what is around, taking us to an Aleph without having to descend the nineteen steps of Borges' tale. From there we can see everything. The images we form in the brain are immediately pop and liquid at the same time. They are naturally so, like those strange shapes in the clouds chasing each other in the sky, sung by De André, with those figures that children see best, like little Irene, three years old, my granddaughter, Claudia and Francesco's daughter.

© Casa Editrice Ulrico Hoepli S.p.A. 2022
A. Diaspro, *Expedition into the Nanoworld*,
https://doi.org/10.1007/978-3-030-94472-8

Are you ready to step into "Pop Microscopy" and let go of that sense of wonder that can arise when colourful imagery brings back memories of something personal? Axl Rose from Guns N' Roses stands out among the green of Ireland, with its red foliage, to become the roots of a tree of Guccini's memory. Ophelia will give way to the Little Mermaid, while a night skiing in Val Badia, thinking of Ivan and Simona, will bring to mind the landscape of La Villa, between Colfosco and San Cassiano. Here we go on tiptoe, to bring curious hearts and brilliant minds closer to science through Pop Art when everyday objects became icons, Andy Wahrol's Campbell's soup can being one of them. The titles chosen for "Pop Microscopy" are part of a game between father and daughter that mixes references from everyday life, from songs to TV series and video games. May the "tourists" of microscopy let themselves be influenced by scientific images, free their imagination and reach us in the Aleph. Everyone should push themselves to give the image that attracts them the title they prefer and then go and find out what it really is, between the technique used and the social motivation. "Pop Microscopy" as a prelude to the "different light" between electromagnetic waves and photons that is the new frontier of optical microscopy, made of liquid images and multi-messenger signals. Here are the images, with their stories under the microscope between a piece of glass and the rainbow.

The story through images is here for you.

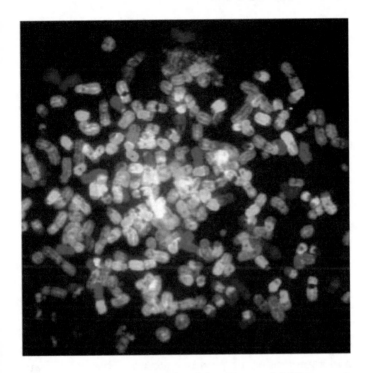

The scientist, image realized by George McNamara: it is the spectral karyotype of the metaphase of chromosomal diffusion of abreast cancer cell, who knows why Claudia thought of the image of a scientist.

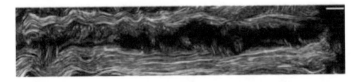

Ofelia, by Chiara Peres, Paolo Bianchini, Alberto Diaspro: it is the meniscus of a bovine knee made without contrast media, *label-free*, exploiting the second harmonic signal. The lake and Ophelia are an almost scholastic memory.

Lana, by Dario Donnarumma: confocal image of a cotton fabric used as a mofo system to study the buoyancy and transpiration of porous materials. That ball of yarn that Grandma Carmen uses for the covers of Claudia and Irene and Teresa has to put back.

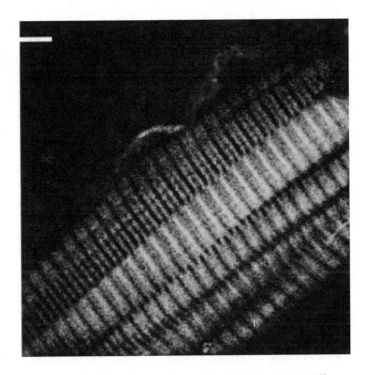

La Sirena, by Chiara Peres, Paolo Bianchini, Alberto Diaspro: they are fillings of myosin from "zebrafish" embryo to develop endoscopic systems. A film that can not be missed between New York, Daryl Hannah and? Neil Young.

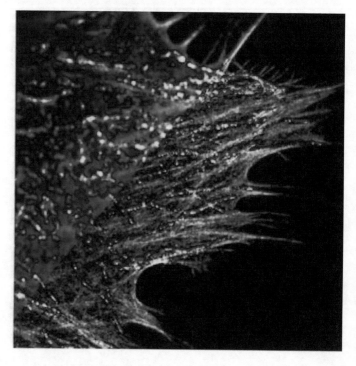

A leaf, by Paolo Bianchini, Alberto Diaspro: it is the cytoskeleton of a cell seen with confocal microscope and fluorescence super solution STED microscope with the aim to understand the role of cytoskeleton in neurodegenerative or oncological mechanisms. That dewdrop that makes the ribs look better is something Claudia has known since she was a child.

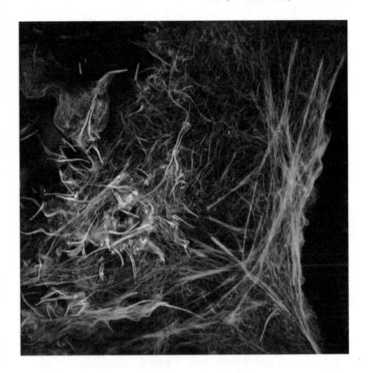

Glycine (Wisteria), by Eric Betzig, Nobel Laureate in Chemistry in 2014: these are actin and alpha-actin filaments captured using in non-linear and linear mode the super-resolution devised by Mats Gustafsson in structured light, SIM. The image Eric sent came out of the electronic message scented.

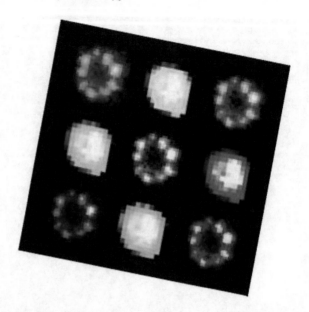

Pacman, by Stefan W. Hell, 2014 Nobel laureate in chemistry: it sounds like a game but it's details, protein Cep 164, of eukaryotic cell centriols visualized at 50 nm resolution with STED and crucial in cell reproduction in health and disease. It is a game that several generations share as Stefan W. Hell's ideas and obsession for spatial resolution with the light microscope.

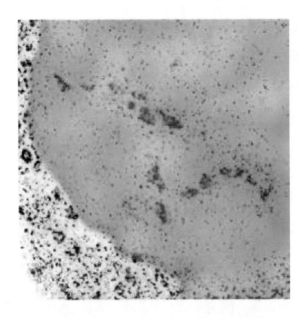

Antigua, by Paolo Bianchini, Gabriella Viero, Alberto Diaspro: it looks like a pirate map taken from the tales of Emilio Salgari and Daniel Defoe, while it concerns the incredible spatial organization of the ribosome machinery that earned Ada Yonath the Nobel Prize, captured in correlative mode with STED.

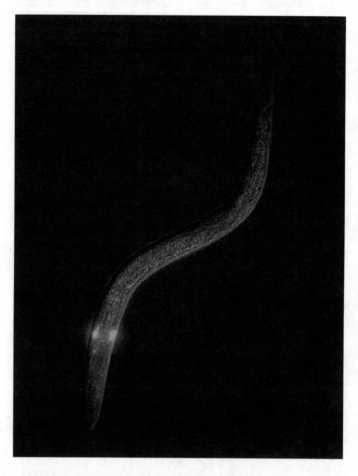

The Snake, by Martin Chalfie, 2008 Nobel laureate in chemistry: green fluorescent protein is made to express, produce, in the living to visualize the development of a nervous system or cancer. Here we are dealing with neurons from *C. elegans*, a nematode, a model system for understanding how humans function.

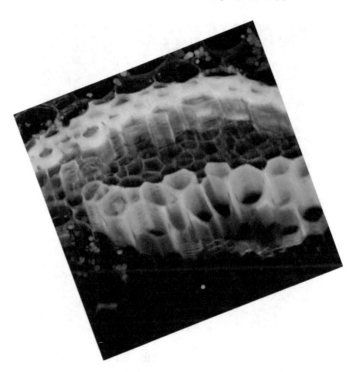

Equator, by Paolo Bianchini, Gabriella Viero, Alberto Diaspro: it is not Pino Daniele's *Penguin at the Equator* but a three-dimensional image that exploits, without the need for marking, the intrinsic fluorescence useful for studying the mechanisms by which normal cells become cancer cells.

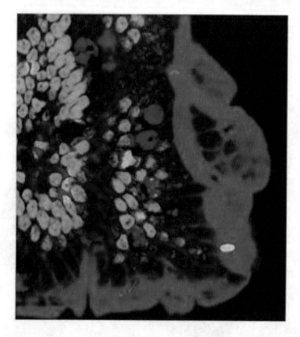

Magnolia, by Paolo Bianchini, Alberto Diaspro: it's the multi-fluorescence are the multimessenger signals sent by cells of mouse intestine, which are due to two-photon excitation. So beautiful is this image, used by many to represent microscopy, that it could only evoke the magnolia with its alternate, oval or elliptical leaves and those solitary and very large multicoloured flowers.

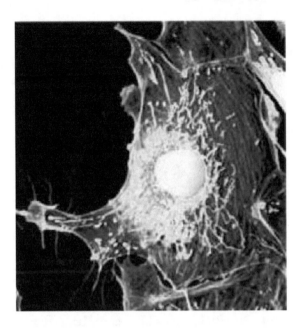

Sperduto, by Alberto Diaspro, Mirko Corosu: it is multi pla fluorescence of epithelial cells obtained with the first two-photon microscope built in Italy thanks to the National Institute for the Physics of Matter back in 1999. Thanks to Manuela Arata, Cristiana Ricci and Mauro Robello for having believed in that project done with Nikon's first confocal head, the PCM2000, at the Physics Department of the University of Genoa.

A frog, by Paolo Bianchini and Paola Ramoino at DiasproLab, leaves of *Lemna minor* imaged by a confocal microscope. Claudia had in mind a Sega-Gremlin video game developed in 1978. Ave have you ever played *Frogs*, a video game action game, developed by Se- developed by Sega-Gremlin in 1978, with your daughter? It's the first video game with a jumping character and the player jumping around controls a frog and tries to catch it with his tongue of the animal and as it jumps various insects. I think that's what Claudia came up with.

Radici (dedicated to Francesco Guccini), by Chiara Peres, Paolo Bianchini at Diaspro Lab: it is the planar section of bovine meniscus with second-harmonic microscopy using polarized light; the dedication to Francesco Guccini in the idea of overcoming the need for biopsies without the need for contrast media is unmissable. Francesco Guccini has been for me and Teresa and then for our daughter Claudia the soundtrack of our whole life. It still is.

La Ila (dedicated to Val Badia), by Michele Oneto, Paolo Bianchini at the Diaspro Lab: it is the multiple confocal colour microscopy of mouse retina cells and it reminds Claudia and me of the places that my grandparents, in about 1936, explored with an old Ardea. Since then the Diaspro always go there, between the Hotel Riposo, the Hotel Mezdì, the Ciasa Soreghina and the never left Tabladel of Ivan and Simona. "La Ila" is the locality of "La Villa" seen while skiing down from the mountains at night.

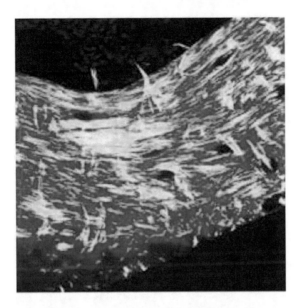

Axel, by Paolo Bianchini at Diaspro Lab: it is artificial bone studied with microscopy without contrast media, the second harmonic. For Paolo it's bone, for Claudia "the hair" of Guns N' Roses.

Green sleeves, by Thankiah Sudhaharane, Jaron Liu: these are the protrusions induced in neuroblastoma cells view with su- microscopy per-SIM resolution, structured light so dear to Mats Gustafsson; for Claudia and me they are the notes of *Greensleeves*, a famous folk melody of English tradition, the first guitar practice in Franco Cerri's course.

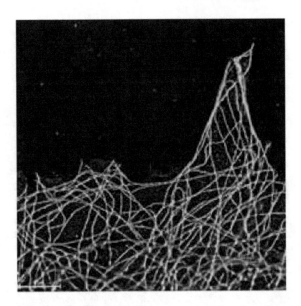

Luna Park, by Francesca Cella Zanacchi, Francesca Pennacchietti, Alberto Diaspro: it is a 3D cellular network in single molecule microscopy visualization realized with Nikon's N-STORM microscope; it is part of the history of single molecule visualization of the Diaspro Lab, from the Physics Department to the Italian Institute of Technology and before that to the Biophysics Institute of the Faculty of Medicine of the Genoa Ateneo. Places that Claudia experienced as a child, as an adolescent and as a woman, with the Luna Park of the Foce never too far away.

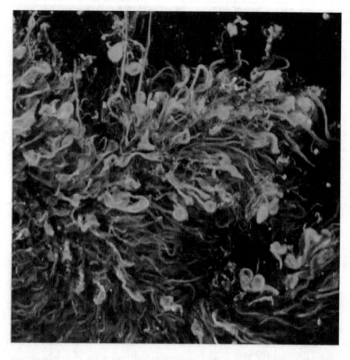

Ireland, by Michele Oneto, Paolo Bianchini at Diaspro Lab: it is a two-colour confocal microscopy of retinal rods, an important study in the understanding of the mechanisms of vision. Claudia, who was working on these images, was then waiting for that creature who, if a daughter, would be called Irene. This is Ire Land, Irene's land because of its beauty and all that green around it.

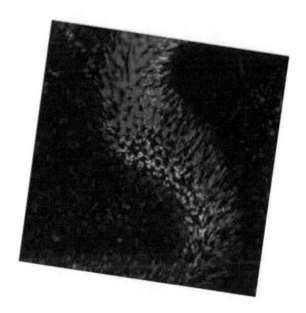

Ariel, by Paolo Bianchini at Diaspro Lab: confocal laser microscopy bovine retinal rods with three-dimensional space encoded in colour. This is pure Disney for Claudia, this is *The Little Mermaid*.

This travelling exhibition has been at the Festival of Science in Genoa, at the Fine Arts in Carrara, at the Year of Light in Parma, at the "G. Doria" Museum of Natural History in Genoa, at other popular events and has represented a piece of Italy, with Ilaria Capua, at the University of Florida and at the celebrations in Washington of the five hundredth anniversary of Leonardo da Vinci.

Printed in the United States
by Baker & Taylor Publisher Services